FIRST PUB
ON THE RIGHT

FIRST PUB
ON THE RIGHT

A Wife-Changing Motorcycle Adventure
from Cork to Cape Town

David Irish Anderson

 PURE INK PRESS

PURE INK PRESS

Paperback ISBN: 979-8-9875866-3-1
Ebook ISBN: 979-8-9875866-7-9

Library of Congress Control Number: 2024924525

www.pureinkpress.com

Contents

For Phyll & Noel.
Always looking twice.

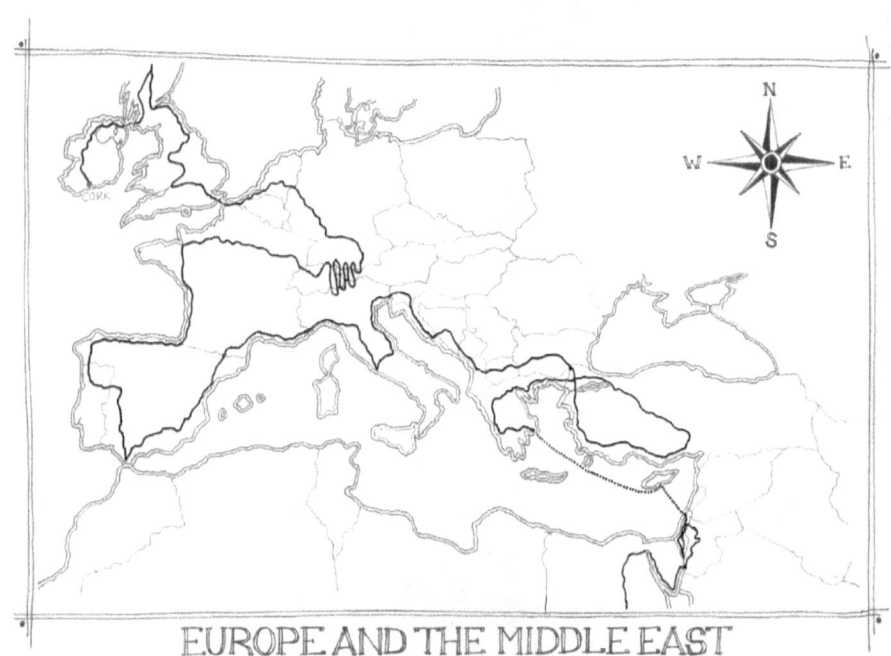

EUROPE AND THE MIDDLE EAST

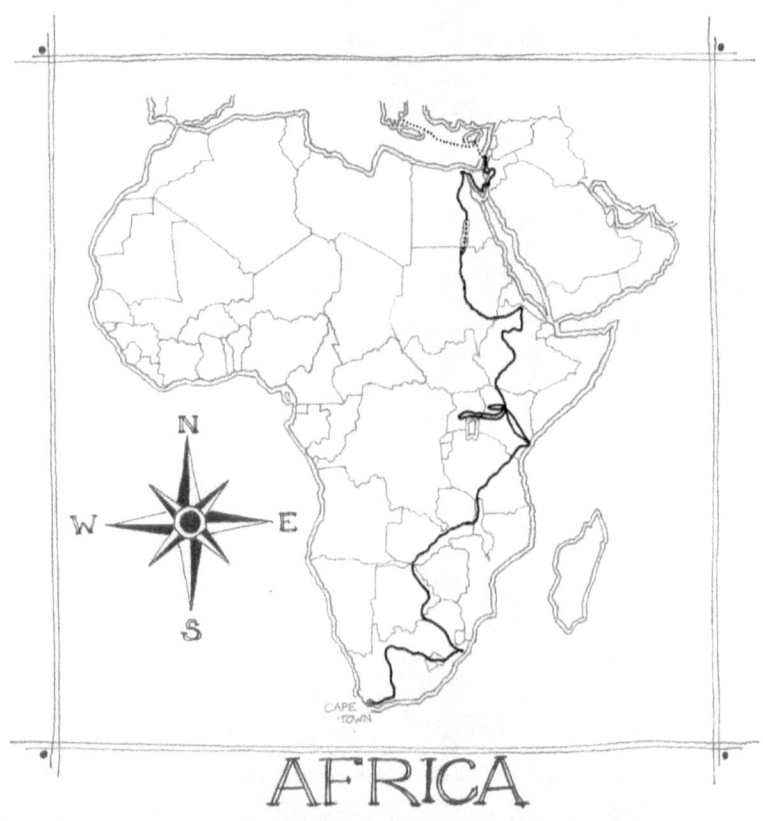

AFRICA

Letter to the Reader

Dear Reader,

This book is more than a chronicle of miles travelled and destinations reached; it is a deeply personal account of the joys, heartbreak, and transformative moments that shaped the person I have become. It is a testament to the unpredictability of life and the power of the human spirit to adapt, endure, and ultimately find peace amid chaos.

Thank you for joining me on this journey through my memoir. I hope you find as much joy, reflection, and insight in these pages as I did while living and recounting them. Through the highs and lows, the laughter and tears, I hope to share with you the lessons I learned along the way. This story is about the nature of forgiveness, the strength found in vulnerability, and the beauty of discovering one's true self. I hope it offers you some comfort, a little wisdom, and a sense of our shared humanity.

Please note that to protect the privacy of those I encountered along the way, certain identities and details have been changed. Making these modifications out of respect for the individuals who crossed my path does not alter the essence or truth of our shared experiences.

With gratitude and hope,

David "Irish" Anderson

CHAPTER ONE

Thirsty for More

The seeds of this adventure took root one evening in 2007 in the cosy confines of a small pub on the right-hand side of a sleepy little road in the Colorado mountain town of Minturn. I remember little else about that auspicious night. It was the beginning of yet another northern hemisphere summer, and my fellow river guides were returning from around the world, eager to begin a fresh whitewater rafting season.

Buzzing from a successful day on the bone-chilling Eagle River, we assembled in the comfy surrounds of the Minturn Saloon to share tales of our exploits over the previous months. The beer flowed like wine, as did shots of cheap whiskey and tequila served in thin paper cups. The sound of excited conversations gradually rose to a deafening crescendo until we had to shout to be heard. Word of an impending celebration quickly spread, and the bar was soon filled with familiar faces, easy smiles, and sparkling eyes. As the night wore on, pockets of the like-minded among us began forming conspiratorial clusters around the heavy wooden tables.

My wife and I had just returned from a thrilling rafting season in South America, and our unusual method for getting there was attracting plenty of interest. With a few months of downtime between the northern and southern hemisphere seasons, we had decided to ride from Colorado to Chile on the back of a single motorcycle.

Sarah and I met in Northern California where we both worked as whitewater river guides. This memorable moment happened in the beer garden of the rowdy Coloma Club, a raucous bar in the heart of a town that thrived on its vibrant rafting community. It was a hot and humid evening when she first approached me and asked if I wanted to "make out." I told her I had no idea what that meant.

"I think you'll like it," she whispered in my ear, her soft breath tickling the skin on the side of my neck. She took my hand, led me to a secluded corner, and showed me exactly what she meant – then she vanished as quickly as she had appeared. She was right – I liked it, I liked it a lot.

Over the next few days, I asked around to uncover the identity of this mysterious woman who left me intrigued and wanting more. To my dismay, I was told she was already involved with someone. Disheartened, I backed off and lost some respect for her.

A week later, she tracked me down, asking why I hadn't sought her out.

I told her what I'd heard and explained that I wasn't interested if she was already seeing someone else.

She let out a beautiful, melodious laugh, her eyes twinkling with amusement. "Yes, I'm seeing someone," she said when she could catch a breath. "It's *you*, dummy!"

We somehow made it work with our hectic lifestyles and eventually eloped to South Lake Tahoe three years later, cementing our connection in a spontaneous, love-filled ceremony. The motorcycle trip to Chile served as our unconventional honeymoon.

Even before we had departed for Chile, friends and family were warning us of the dangers of travelling through Central and South America. Newspaper articles and TV reports painted a bleak picture of the lands we would cross, so the adventure began with a significant amount of trepidation.

Our ride through Mexico caused us to reassess many of our prejudices and expectations. When the locals constantly met us with compassion and acceptance, we felt humbled and, at times, ashamed.

As we neared the border between Mexico and Guatemala, we began to hear familiar warnings again. Only this time, the Mexican people were telling us stories about those in Guatemala. It was a pattern we observed many times throughout that journey – people of the country we were leaving would warn us about the people of the country we were about to enter. And yet none of those warnings matched any of our experiences. Warm welcomes, kindness, and support followed us throughout the entire voyage. Before we knew it, we had arrived at our destination in Santiago, Chile, approximately two and a half months after leaving Colorado.

As Sarah and I retold stories from along the way, several friends expressed interest in doing something similar on an even grander scale. The whiskey and tequila shots were clearly starting to take effect as some of the suggested routes would involve building roads where none currently existed. Others were so ambitious they would have taken many years to complete. I began to question some of my friends' knowledge of the world beyond the borders of the USA. As the night wore on, we finally hatched a plan for several of us to ride a fleet of motorcycles from Cork, Ireland, to Cape Town, South Africa. And so the idea for "Cork to Cape" was born.

The original six instigators included an eclectic mixture of whitewater rafting guides from four continents. As the evening came to a close, we had all become infected with a contagious excitement for the adventures to come, even though it was clear from the beginning that such an undertaking would take months, if not years, to plan and prepare for.

Life is something that happens while you are making plans. As time passed, one by one, each of the people who had initially expressed an interest in being a part of this adventure found alternative paths, obligations, and commitments until only two remained: Sarah, and me. Since our motorcycle trip to Chile, we had both been itching to get back on the road for something a little more intense. But this time, on two motorcycles.

Our shared passion for whitewater had taken us around the world. Unfortunately, the guiding industry is notorious for underpaying and undervaluing the people who work in it, so a year after we returned from Chile, Sarah felt drawn towards finding a career with more stability and better remuneration. With a lot of hard work and determination, she trained and qualified as a registered nurse. Shortly afterwards, she found a job at a large hospital in California. I also reluctantly drifted away from the rafting industry and secured a position as project leader with a construction crew to help subsidise our income and provide the funds necessary for our expedition.

***

I don't know when my interest in travel first emerged, but I still have vivid memories from my childhood of staring for hours at a large map of the world that adorned the wall, opposite my seat at the family dining table. That map provided a welcome distraction from the soggy brussels sprouts I would push around my dinner plate in the hope they would magically disappear.

I grew up in a small town in Northern Ireland called Strabane during a period known as "The Troubles." It always struck me as a rather odd thing to call what was, in essence, a war – an irregular war indeed, but a war nonetheless. The British and their nomenclatures would have called it "The Mild Inconvenience" or "The Kerfuffle" if they could have gotten away with it.

My childhood coincided with a time of bitter conflict.

Political, religious, and socio-economic differences erupted into an armed struggle, the roots of which could be traced back over hundreds of years. Despite the violence and strife, we were just ordinary people living through extraordinary times. Amid the chaos, one of the few relatively stable spaces was provided by the education system.

Throughout my youth, I assumed it was normal to see tanks and soldiers in the streets, military helicopters overhead, and sporadic bombings and shootings. My first taste of an alternative world came at the age of thirteen when my high school geography teacher made the decision to show our class a film about him canoeing down the wild rivers of Norway. As his dusty, old projector chattered into life, it cast ghostly images onto a large white sheet draped over the classroom blackboard. A typically rowdy class of teenage boys sat silently entranced as we watched him navigate what looked like impossible whitewater rapids. I was immediately hooked on the idea of learning these mysterious skills and following in his footsteps, or rather, paddle strokes.

Under that teacher's guidance, a group of my peers and I helped establish a tremendously active outdoor pursuits club within the school. Before long, we were getting lost in the mountains, falling off rock faces, and swimming after our kayaks down many of the local waterways. Under his tutelage, we roamed far and wide in search of the next challenge and escapade. My passion for adventure was conceived and nourished by a single teacher willing to sacrifice his weekends to share his love for the outdoors.

I was instantly inspired to make this my path, too.

Before long, I had met professionals within the outdoor education field who dedicated their lives to doing the same. After school, I moved to Glasgow, Scotland, to pursue a degree in outdoor education. Scottish mountains were taller, and the rivers were steeper. I spent every spare moment I had between my studies exploring the rugged highlands with a dedicated band of enthusiasts from my university program. Around this time, I first sat in a whitewater raft and realised I could take many people together down the rivers I loved so much.

Early in my career as a river guide, I decided to make it my mission to work on every continent where rafting was practised. Over the next three decades, I travelled the world, chasing rivers and sharing adventures with as many people as possible; from Europe to Africa, then to Asia, New Zealand, North America, and finally, South America. When I had achieved that goal, I felt somewhat directionless. After the evening at the Minturn Saloon, the idea of riding from Ireland to South Africa became an obsession for me. The thought of riding from Cork to Cape Town began to fill that hole in my life.

It was a chance encounter with a Chilean guide while working in New Zealand that had inspired the motorcycle trip from Colorado to Chile. Sebastian Astorga, manager of *Cascadas de las Animas*, a rafting company based on the Maipo River outside of Santiago, invited Sarah and me to go and work for him the following year. The release of the movie *The Motorcycle Diaries,* documenting Che Guevara and Alberto Granado's motorcycle adventure across South America, also happened around this time.[1] That gave us an idea.

Why not ride a motorcycle across the Americas to get to our new home and jobs?

We got busy planning. After completing another northern hemisphere rafting season, saving every spare penny for the equipment necessary for such an undertaking, and reading numerous travel guides, we saddled up our BMW R 1100 GS in Colorado to begin our first long-distance overland motorcycle adventure.

I had promised Sebastian we would arrive in *Cajon del Maipo* on the 14th of November, 2006. After two and a half months on the road with more than our fair share of wrong turns, upsets, and motorcycle problems, we rolled into his rafting base at 2:30 p.m. on the agreed-upon date. We were as surprised as he was that we arrived on schedule; and yet, with such a journey behind us, we wished we'd made more time to stop, explore, and absorb the beautiful regions we had ridden through. We were thirsty for more.

Even though I had always shied away from fixed itineraries, I knew a journey of this magnitude, from Cork to Cape Town, was going to require more thorough preparation and detailed planning.

It would take Sarah and me six years.

We spent these years preparing, saving, and acquiring everything we thought we might need for the journey. I'd once met a German adventure rider who gave me this advice: "Once you get your bike and gear, point your machine in the direction you want to go and open the throttle – everything else will work itself out." But with the number of borders we would have to cross, along with the visas required for each, we felt compelled to investigate what restrictions or difficulties we could possibly encounter. Not all passports were welcomed in all the countries we intended to travel through. I would be travelling on an Irish passport, while Sarah would use her US one, so the North

African route from Morocco to Egypt via Libya would not be viable. Our eventual plan would take us through Syria, but by 2013, with a civil uprising escalating into a full-scale war, this option was beginning to look less attractive.

I had travelled through war zones before – I grew up in one. I learned that the conflict often takes place in very localised areas. There were often entire regions left completely untouched and unaffected. I didn't want to take the Syrian option off the table until we got closer to the situation.

Our previous experience on a single BMW motorcycle impressed upon us their reliability and durability. This time, we would be on two bikes, Sarah bravely insisting she wanted to ride her own machine. Her shorter stature made many of the taller enduro motorcycles a little too intimidating.

To limit the number of spare parts and tools we would need to carry, we agreed that similar machines would be the wisest choice. We finally settled on the BMW F series. I would ride the 800 GS and Sarah would ride the slightly lower 650 GS. The bikes were similar enough that many of the components would be compatible and interchangeable. It seemed like the logical choice, if a little heavy and cumbersome for some of the conditions we imagined we would face. We eventually found two bikes on the used motorcycle market; each with reasonably low mileage and in near-perfect condition.

Next came the task of modifying and outfitting each one with luggage options, camping arrangements, cooking equipment, clothing, and tools. With limited space, every piece of the puzzle had to be compact and lightweight. The tent, our primary living space, had to be just right – big enough for us and all our stuff, yet small enough to pack inside our panniers. After much deliberation, we settled on a bright green three-person dome tent with an extended vestibule. We chose a camping stove that used petrol so we could share fuel between it and the bikes. Our sleeping pads were inflatable mattresses that compressed to the size of a pint glass. They appeared too delicate to provide a comfortable night's sleep, so we tested them in very rough terrain, and they never let us down. We could zip our mummy-style sleeping bags separately, or to each other for snuggling in colder conditions. As a novice guitar player, I was determined to find a way to carry one with me. The odd shape, fragile neck, and sensitive tuning dissuaded me from taking my precious instrument. I was almost ready to abandon the idea when I found a manufacturer that produced a foldable full-size guitar. The dimensions, when folded, fit perfectly inside a rigid plastic Pelican box. With some modification, I created a durable case that I strapped securely onto the back seat of my machine. Solving this challenge brought me more joy than I thought was possible. On top of purchasing the necessary gear, we would also have to cover the day-to-day expenses. It was becoming apparent that it was going to be a costly undertaking.

Our previous adventure had taught us we could get by on approximately twenty British pounds – roughly thirty US dollars – per person, per day. Although we had no fixed agenda, we figured we could be on the road for one to two years. We knew that while Europe would be relatively expensive, Africa would be significantly cheaper, so we felt comfortable with our estimated average. This amount would have to cover fuel, food, accommodation, visas, motorcycle upkeep and repairs, and any other unexpected expenses we would encounter. If it sounds tight, that's because it was, but years living on a guide's meagre income made us resourceful with the skills and discipline to make it work.

We set the date of our departure for May of 2013. Being six years out, it felt like we had all the time in the world. We covered the walls of our modest apartment with massive maps of Europe, the Middle East, and Northern and Southern Africa; we would highlight points of interest as our planning progressed until a rough route emerged. With each day that passed, we felt the anticipation growing.

On more than one occasion, we questioned the wisdom of what we were doing. We'd started to become comfortable and somewhat settled in our routines. Sarah's well-paid nursing position at a nearby hospital and my construction and truck-driving jobs proved quite lucrative. From all appearances, it looked as though we were living the dream.

Walking away from it all seemed risky and irresponsible. Yet the hunger for another adventure kept us on the move, and before we knew it, we were riding our loaded motorcycles into a shipping depot on the outskirts of Sacramento.

That was the point of no return.

We quickly and efficiently stripped the bikes, removing wheels, handlebars, and mirrors so they would be as compact as we could make them. Strapping them both onto wooden shipping pallets, we performed a final inspection. We bade them farewell as they began their month-long overseas voyage to Ireland – where we would start our adventure.

As we left the shipping depot, the sun was setting, casting a warm glow over the city. The air was cool, and the streets of Sacramento were quiet except for the occasional rumble of a passing vehicle. We couldn't help but feel a sense of gratitude for the opportunity to pursue our dreams and explore the world in our own way.

The road ahead was uncertain, but we felt ready to face whatever challenges we would meet. While we knew we would encounter obstacles, setbacks, and unexpected twists and turns, we were determined to overcome them and keep moving. Our spirits were high as we focused our minds on the adventure that awaited us.

As our taxi drove into the night, the stars above twinkled like tiny diamonds in the sky, reminding us of the vastness of the universe and the limitless possibilities. We were just two adventurers beginning our travels; yet, in that moment, we felt part of something much bigger than ourselves. We were a part of the universe, and the universe was a part of us.

Suddenly, the realisation dawned on us that, after six years of planning, it was actually coming together, and our odyssey had unmistakably commenced. Echoes from that night at the Minturn Saloon, when a group of us had first conceived this wild idea, drifted back to us. Laughter over clinking glasses and the bold promises made beneath soft lights transformed into the reality we were about to embrace.

CHAPTER TWO

Afraid of Not Living

Personal Diary Entry: 18th May 2013
Day 5 (of 549)
872 km of 62,840 km (541 mi of 39,047 mi)

"Go south, my boy," my father whispered to me as he bade me farewell. With a reassuring pat on the back, he sent me on a journey that would change my life forever.

So, it was finally underway. The Cork to Cape Town adventure had begun.

The final couple of weeks prior to leaving had been a rollercoaster of red tape and false starts, but it now felt as though we were making progress. Sarah's parents were kind enough to let us use their home in California as a staging area for some last-minute preparations before our departure, and they threw a farewell party for our last night in the Sierra Nevada mountains. With a fixed deadline and pre-arranged flights, there was no turning back. It was time to move on.

A smooth flight across the Atlantic took us directly into Dublin. Shortly after landing, we cruised north in a sporty little rental car towards my hometown of Strabane. It had been almost two years since I'd set foot on the "auld sod," and the anticipation of returning had kept me awake for the entire journey. As the roads became more familiar and the landscape of my childhood unfolded before us, I felt deeply comforted. Even the unruly Irish weather blessed us with a rare sunny day.

Lush green fields blanketed the rolling hillsides surrounding the town I had left behind in my youth. My father greeted us with warm hugs and endless cups of tea as the unique aromas of home triggered wave after wave of memories I thought I had long forgotten. As news spread of our arrival, family, friends, and neighbours gathered, turning our homecoming into an excuse for a lively celebration.

With a fixed base for a few days, we had the time and resources to track down our motorcycles, which were scheduled to arrive in Dublin later that week. Unbeknown to us, however, while we travelled to the starting point of our trip, a US Customs official in Texas was fastidiously making a mess of our shipping arrangements, assuming we were exporting our bikes with the intention of selling them in Ireland. With Irish and US Customs now involved in the process, what should have been a simple shipment became a bureaucratic nightmare, with multiple agencies demanding their pound of flesh for the importation of our own motorcycles. Making no progress over the phone, we decided to drive to Dublin and talk to the parties involved in person to explain our situation. We eventually managed to convince an Irish Customs inspector that we were not planning to permanently import our motorcycles. A week behind schedule, they finally granted us permission to drive our reassembled bikes out of a bonded warehouse in Swords, Dublin.

It felt a little surreal.

We rode our motorcycles south towards Cork, beginning a voyage that had been no more than a dream for so long, but there we were: my father and my brother, Michael, both of whom had vowed to join us on their bikes for the Irish leg of our ride; and Sarah and me on our machines. The four of us drove into the sunset at the end of a productive but exhausting day.

On what was to be the first official day of our adventure, we rose early at our B&B, packed the bikes, and made our way to one of Cork's most recognisable landmarks, Saint Anne's Church, known for the Bells of Shandon. The locals called it the "Four-Faced Liar" – each of the clock tower's four faces tells a different time – it served as a fitting starting point for the Cork to Cape Town ride.

My father kickstarted his 1950 BSA 123cc Bantam into life with a loud backfire and a cloud of smoke. It was the same model of bike he had ridden as a young man and had a top speed of 50 kph – going downhill with a tailwind. He had found this one in England a year before, and had spent months lovingly restoring it so he could join us for the beginning of our journey. My brother pulled up on his Kawasaki ER-6F behind us, and Sarah and I whooped at the front of our little convoy as we began our ride north.

We spent the next few days winding our way up the West Coast of Ireland, enjoying the tight, twisting country roads while experiencing all kinds of weather conditions. However, with the delay at Dublin port, we were on a much tighter schedule than we had planned. What was supposed to have been a relaxed exploration of the West Coast became a race against the clock. We had timed our arrival in Ireland to coincide with the beginning of the North West 200 road races, the local equivalent of the Isle of Man Tourist Trophy – arguably the world's most famous and dangerous motorcycle race. Many of the TT riders were now gathering in Ireland to test their skills and nerves against each other and the challenging course, a fourteen-kilometre street circuit known as the Triangle, situated between the towns of Portstewart, Coleraine, and Portrush on the north coast.

On Thursday, we made it to Portstewart just in time for the opening evening of practice sessions. Thankfully, we got to catch three of the races that evening because the torrential rain had other plans for the rest of the event. We braved the elements on Saturday morning and stood with thousands of spectators along the impossibly narrow roads of the treacherous course. The fearless riders pushed their machines to their very limits, each one a blur of mist and noise as they screamed past us, but the weather was to be the outright victor on that day; the marshals called off the event after several high-speed crashes. Though disappointing, it was obviously the outcome of the logical reasoning that needed to happen in such scenarios.

I needed a break from logic and reasoning.

We drove to Belfast to stay with my sister, Joan, who worked there as a secondary school teacher. Before we left the island, Sarah and I promised Joan we'd have a little visit with her young students.

Early Monday morning, when we drove our fully-loaded motorcycles onto the Jordanstown School grounds, an excited group of curious teenagers greeted us, chattering incessantly as they led us to their classroom. In the packed room, everyone gave us their full attention as we outlined our trip and fielded questions about our hopes and plans.

"Aren't you afraid of dying?" one of the children asked.

"I think I'm more afraid of not living," I responded.

That afternoon, we met up with my younger sister, Gerrie, at a bar she managed in Belfast. After a late night of comfort food and beer, Sarah and I tumbled into bed tired but electrified with the thrill of knowing we were about to leave Ireland and enter the second country we'd ride through on our trip.

When the morning's sunshine slowly dissolved the soothing darkness, we dragged ourselves out of bed and hurried to catch the boat to Cairnryan in Scotland. The slow ferry across the North Channel gave us ample time to contemplate the whirlwind of encounters with family and friends, a blur of "hellos" and "goodbyes" marking the beginning of our journey and the end of its first leg. From this point forth, it would be just Sarah and me.

This trip had been only an idea for so long, and it was still hard to believe we were on our way.

As we docked in Scotland, we quietly mounted our bikes inside the damp, dark belly of the ageing ferry before taking the less-travelled coastal road towards Glasgow and beyond. We stopped frequently along the way to enjoy the stunning views and the rare fine weather. An overnight stop at the foot of Ben Nevis, Britain's tallest mountain, left us wondering if we could fit in a hike and still make it to our destination of Findhorn at a reasonable time. We made it work, rising early the following morning before hoofing it up the Ben in record time. We were back on the road by early afternoon, getting into Findhorn in time to watch the sunset over the bay as we munched on the Scottish staple of fish and chips. With each mouthful of delicious, salty greasiness, I was sure I could feel my heart rate slowing and my arteries thickening.

Turning south and driving into England, we made a short stop at the Holy Island of Lindisfarne. We took our bikes across the tidal causeway connecting the monastery to the mainland for more than a glimpse of the ruins from AD 635 and the 16th-century castle.

We went to the Holy Island for the wine.

As the monks poured us samples of their famous honey wine, I playfully elbowed Sarah with a mischievous look. When she didn't get my message, I leaned over to whisper that I assumed the red-faced, jolly-looking monks had been doing plenty of tasting themselves as they shuffled around in their traditional tunics. There had to be some perks to living on a remote island after taking a lifelong vow of celibacy. It appeared that even in the realm of devotion, a touch of mirth and earthly pleasures found their place.

With a rising tide threatening to cut off our road back to the mainland, we reluctantly bade the mead-merry monks farewell before continuing along the coast to Whitby and through to Robin Hood's Bay, and as we navigated the winding coastal roads, a picturesque tapestry unfurled. Quiet villages nestled along the way, their charm highlighted by the backdrop of rolling farmland that stretched off as far as the eye could see, and majestic hilltops showcased weathered castles, their presence a testament to a bygone era. With each twist and turn in the road, we became immersed in the beauty of a landscape where history and natural splendour coalesced into an unforgettable journey.

When planning our trip, we knew this region of England would be too charming for us not to explore.

So we'd allocated some of our time to allow us a comfortable stay at a B&B – before we even deserved such a break. Northcliff Bed and Breakfast, situated within a five-minute walk from Robin Hood's Bay, was run by Callum, a partner-in-crime from my college days in Scotland. Robin Hood's Bay, resembling an M.C. Escher lithograph come to life, is a unique settlement where quaint, little cottages cling impossibly to the precipitous coastline within a warren of steep paths and secret passageways. Callum's boisterous children proudly served as our impromptu tour guides as we journeyed through their beloved hometown. Their youthful enthusiasm breathed life into the tales of infamous pirates and hidden treasures, vividly portraying the area's rich maritime history. Reluctantly contemplating our departure, we longed for the luxury of another day in this enchanting location.

It was only the beginning, and it was time to move on.

With a nasty storm chasing us from the north, we rode south to Donnington Park to attend the 2013 Horizons Unlimited (HU) UK meeting. Conceived in 1999, HU was a forum for overland travellers wanting to share current information and advice from the road. It rapidly grew into a community of individuals from every corner of the planet. Sarah and I became active members after our journey through Central and South America. Part of our commitment to the HU community involved offering assistance to people travelling through our region. On top of the wealth of relevant information HU provided, it also organised regular meetings where it was good to rub shoulders with like-minded souls who didn't frown upon us for what we had set out to do or question our sanity for undertaking such an adventure. Indeed, our trip paled compared to some of the itineraries planned by our fellow travellers.

For the next four days, dozens of adventurers from around the world gathered at a converted farm to share tales and advice from the road. Workshops and presentations covered a variety of topics. There was even a class on how to prepare and cook roadkill! The onsite toilet facilities appeared busier on the morning after that event. A challenging motorcycle obstacle course was set up on the grounds to test our off-road skills, and professional photographers came by to show off their ability to capture a clear shot of someone zooming past them at high speed. Established travel writers gave advice on crafting engaging stories from the road while budding filmmakers debuted their latest work. By the end of it all, we felt inspired and ready to take on anything our journey might throw at us.

Soon, we were pulling into Bristol, where we had more family to visit. During those few enjoyable days, we spent a large part of one day hunting for the elusive murals of local street artist, Banksy, hidden throughout the city. As we left on the afternoon of June 5th and made our way east towards London, I realised it would be the last time I would see anyone from my immediate family for a very long time. We looked ahead to the day when we could repay everyone's kind-heartedness.

To my father, I had made a solemn vow. Every time I gazed upon the wonders that awaited me, I promised him I'd look twice – once for me, and once again for him.

CHAPTER THREE

The Spontaneous Adventure of Being Hopelessly Lost

Personal Diary Entry: 5th June 2013
Day 23 (of 549)
3,661 km of 62,840 km (2,275 mi of 39,047 mi)
Every decision I have made in my life, every opportunity I have taken, every left turn or right turn I have chosen, has brought me to this place and time; and this moment is all that I have.

On entering London, we prepared ourselves for a noisy night in the middle of the concrete jungle.

We had reserved a campsite near the district named after the Crystal Palace that burned to the ground many years before. We didn't expect much – after all, we were in the sprawling suburbs of South London – but we were pleasantly surprised to find a lush camping ground that felt completely removed from the surrounding city.

It felt like we'd stumbled upon a hidden oasis.

A photojournalist we'd met at the Horizons Unlimited gathering had suggested we check out the work of Sebastião Salgado, a Brazilian photographer famous for his breathtaking landscapes and moody portraits, who was currently exhibiting at the Natural History Museum. Hoping to pick up a few pointers on how to improve our prowess with a camera, we left our oasis the next morning and made our way to South Kensington.

Coincidentally, it appeared as though every school in England had chosen that day to take their students on a field trip to the same museum. What we'd hoped would be a peaceful, contemplative experience turned out to be something quite different. After a raucous morning, thoroughly outnumbered, we beat a hasty retreat to the Covent Garden district for lunch.

We spent the rest of the afternoon exploring the West End, somehow stumbling – on more than one occasion – into London's red-light district, before crossing paths with an old Israeli friend from my rafting days in Nepal. Such coincidences delighted me: when I, from Country A, met someone from Country B when we were both working in Country C; and then ran into that same person again in Country D. What a small world indeed.

***

The following morning saw us up bright and early, motoring our way towards Dover for a midday ferry to France, fighting fierce crosswinds along the way. A quick stop for breakfast at a roadside café provided us with a wonderful reminder of the exquisite quality of English cuisine. With full bellies and greasy lips, we rode our bikes into the heart of a cross-channel ferry and sailed to France, saying goodbye to the White Cliffs of Dover, over which there was not so much as a hint of blue sky.

An hour and a half later, after a choppy crossing, we rode onto French soil via the port of Calais. We'd made the decision to push north towards the Netherlands and decided Dunkirk would be as good a place as any to spend the night. After pitching our tent and devouring a delicious camp stove meal, we took a walk along the beach, watching the kite boarders skip across the white-capped waves. It was hard to imagine how it must have looked on those fateful days of the frantic Allied evacuation so many years before.

As a new day dawned, we plotted a course for Bruges, Belgium, to avoid all the major highways. Soon, without even realising it, we crossed yet another border, surprised at how the continent had changed since my youth. Other than subtle differences in the road surface and street signs, there were few other indicators that we had entered another country. Gone were the security checkpoints, passport stamps, customs officials, and currency exchanges – all sacrificed upon the altar of the European Union.

It all stood in stark contrast to the memories I had of growing up in my divided homeland.

My hometown's proximity to the border made crossing it an almost daily occurrence. I would cross from my home to the Republic for various reasons – to visit friends, access great surf on the West Coast, and find cheaper fuel. Each time, I was wary as I approached the border.

Ominous British military compounds straddled every border crossing between Northern Ireland and the Republic. Beneath drab olive gun towers, a maze of concrete barriers prevented vehicles from approaching at speed, so traffic would snake along under the watchful eyes of heavily armed soldiers who nervously aimed their guns at the occupants within. Driving into the Republic was often accompanied by a collective sigh of relief.

It was only in the late '90s when these immense checkpoints began to disappear, a consequence of the peace process.

For obvious reasons, my early experience with border crossings left me with a feeling of dread every time I had to cross a border. On the other hand, travelling around Europe, with its seemingly invisible borders, was a breeze.

How times had changed.

Once we'd crossed from France to Belgium, finding Bruges proved easy, but finding our campsite was a little more complicated. We had deliberately decided to exclude a GPS on this trip as we'd discovered getting lost can sometimes lead to the most fascinating places, geographically and emotionally. But this time, we didn't find ourselves feeling more engaged and connected with our surroundings, and we definitely didn't come across an interesting character emerging from the mist and chuckling as he pointed, "thattaway." We were hungry and tired – in no mood to enjoy the spontaneous adventure of being hopelessly lost. After several hours of frustrating wrong turns, dead ends, and disagreements, we eventually found our campsite.

We hastily set up our tent, cooked a light meal, and then ate wordlessly. We began speaking again as we cleaned up and decided what to do next. Sarah suggested we take a gentle stroll through the campsite. We held hands as we walked, laughing at our endless bickering while lost. Eventually, full and content, we allowed ourselves another silence, this one comfortable and healthy.

I smiled as my mind landed on a memory from long ago.

I hadn't had my driver's licence for very long when a friend asked me for a ride south to Dublin. My father was kind enough to let me borrow his car. Once I had dropped off my buddy, I began the long and lonely drive home. It was late evening when I spotted a hitchhiker heading in the same direction, so I pulled over and offered him a lift. His destination wasn't on my planned route, but it was cold outside, and the lively conversation was a welcome distraction from the monotony of the road. I decided to do more than just help him get closer to where he was going – I would take him all the way there.

After I had dropped my passenger at his home, I realised I was in an unfamiliar part of the country. It was dark by then, and I knew the border was close, so I figured if I just continued north, I would eventually find a crossing back into Northern Ireland. The sky was clear enough to navigate using the stars, but the winding roads made it challenging to stay on course.

Before long, I had absolutely no idea where I was.

Suddenly, the roadside signage changed. Those of the Republic were black on white, while those of Northern Ireland were white on green. It appeared as though I had discovered a secret route into the North devoid of any military presence. For several kilometres, I congratulated myself on my cunning navigational skills before rounding a corner and seeing the distinctive spotlights of a sizable army outpost.

I'd learned from an early age that turning around within sight of the security forces could invoke suspicion and a possible violent response, so I continued weaving my way between the defensive barriers. When I finally reached the young soldier standing in the middle of the road, the headlight beams made the puffs of his breath visible in the chilly night air, and I rolled down my window. I expected the standard list of questions about my intentions but didn't manage to prepare proper answers ahead of time.

"Evening, sir," he said in a youthful voice. He wasn't much older than I was.

"How are you?" I replied.

"So where are you going on this fine evening, mate?"

"I haven't the foggiest idea," I responded.

"You what? Where are you coming from?" was his somewhat puzzled reply.

"I haven't got a clue where I'm coming from, either." I chuckled to myself as I realised how ridiculous I must have sounded.

It took some explaining to alleviate the young soldier's suspicions, but I was eventually able to convince him that I wasn't as clever as I thought I was when it came to finding my way.

I was pulled from my thoughts by Sarah's soft, "Oh!"

She let go of my hand and pointed at two other motorcycles parked nearby bearing US licence plates and two intrepid individuals performing a little routine maintenance. Dawn and Paul quickly became close friends, and we were soon discussing the details of our respective journeys. They were en route to Mongolia and just beginning the European leg of their adventure. We talked late into the evening, sharing any tips and information we thought would be helpful to each other. It was refreshing to encounter kindred spirits on a road that sometimes felt lonely. We rose early the next day to bid them a fond farewell, hoping our paths would cross again soon.

The beautiful city of Bruges, with its faded grandeur steeped in history, had once been one of Europe's most important ports. But the gradual over-silting of its harbour had left it abandoned, destitute, and practically forgotten for over 500 years. Narrow, cobbled streets, an intricate network of canals, grand, open plazas, and elegant architectural wonders all spoke of a past interrupted. Organic street plans from a less-regulated age provided endless surprises for the inquisitive visitor.

We enjoyed an exquisite lunch at the oldest café in Bruges, the Vlissinghe, dating back to 1515, savouring flavours that echoed centuries of tradition. We sipped locally crafted beers from the seemingly endless selection at the Beer Wall. The aroma of freshly baked waffles wafted through the air, enticing us to sample the delectable treats from a street vendor. Sitting on the polished steps of an old townhouse, we relished each bite while the melodic tunes of street musicians enveloped us in a romantic atmosphere.

As darkness descended, a transformation occurred.

Spotlights illuminated the city's landmarks, casting a magical glow on its storied architecture. The bars and cafés came alive, their windows sparkling with warmth and bonhomie. Once forgotten, Bruges shimmered with renewed vitality under the night sky. In this city frozen in time, we witnessed an enduring spirit that whispered secrets from the past while offering moments of delight in the present.

CHAPTER FOUR

Zeb and the Art of Motorcycle Maintenance

Personal Diary Entry: 23rd June 2013

Day 41 (of 549)

5926 km of 62,840 km (3682 mi of 39,047 mi)

With each passing day, I can sense subtle changes in how we both relate to our motorcycles. Bonds are forming between us and our machines. A feeling of deep satisfaction settles upon us when we perform even the simplest routine maintenance.

We were getting used to sitting on our bikes, listening to them hum and roar beneath us, watching the scenery blur around us, and feeling our eyes become razor-focused on the road ahead. And it was in this trance-like state we rode towards the Netherlands.

Rotterdam was to be our next destination. Again, we mostly avoided all the main highways, meandering our way north through sleepy villages and fertile farmland, past windmills old and new, over bridges, under tunnels, and finally approaching the city via Europe's busiest port.

The second-most populated city of the Netherlands proved to be a pleasant surprise. Little remained of the original city centre after the intense bombings of World War II, and Dutch architects had taken it upon themselves to create a lively, tasteful new centre from what was left. An imaginative mix of innovative styles surrounded colourful street markets, giving the city a fresh, young feel – a radical contrast to our previous stop, Bruges. While there was a fierce rivalry between Rotterdam and the capital, Amsterdam, most Dutch people we spoke with quietly favoured the former.

Next up was the city where anything goes.

No visit to the Netherlands would have been complete without a stop in Amsterdam, widely regarded as Europe's most open-minded metropolis. It seemed unfortunate that the relaxed laws on cannabis use and the candid sex industry were what the city was most famous for when it had so much more to offer. Tall, skinny houses leaned haphazardly against each other, overlooking cobblestoned streets all entwined in a spider web of working canals. In this city, the bicycle was king, and crossing the street could become an ordeal of dodging two opposing lanes of cars, trams, bikes, scooters, and hurried locals on foot. The resident population relished the challenge of cycling past at breakneck speeds, deftly avoiding the stoned tourists and treacherous tram tracks.

We took a circuitous route as we entered the metropolis, slowly spiralling our way towards the city centre. We'd found this approach to be a rather useful practice for getting our bearings in new surroundings. The early morning streets were still damp from a heavy dew or a light rain. I could sense a slight lack of traction under my front tyre each time I cornered the bike.

As I accelerated out of one particularly sharp turn, the rear end of my bike suddenly began to fishtail, and I struggled to keep it upright. Once I had my machine under control, I looked in my mirror to see how Sarah had fared, hoping my loss of control had served as an indicator for her to slow down. I let out an audible gasp as I saw her rear tyre lose its grip on the slick road beneath her. She fought bravely, but the weight of her heavily loaded machine was too much, and she went down. She slid across the road on her side, the bike on top of her.

It all seemed to play out in slow motion, and I felt powerless to help.

I pulled over and dismounted, running back towards where she lay prone in the middle of the road, thankful there was light traffic on the quiet street. When I reached her, I knelt beside her, trying to keep my panic from showing.

"Are you okay?" My voice trembled.

"Yes," she grunted. "Just a little shaken. Now, help me get this thing off me."

We assessed her condition once we were back on our feet, with the bike on its side stand. As a nurse, Sarah was well-versed in physiology and quickly established there were no serious injuries. We looked over the bike and, other than a few scratches, nothing appeared to be broken. We hugged each other as the adrenaline began to wear off, and the shaking subsided. I was so relieved she wasn't hurt. The fall served as a sudden reminder that things could go wrong in a heartbeat, and we resolved to be more cautious from this point onwards.

We had agreed upon a camp on the periphery of the city, and after our spill, we decided to take the rest of the day off after setting up our tent. I used the opportunity to lubricate my chain. Unable to fully relax after what happened, I needed a distraction to keep my mind from wandering to all the worst-case scenarios.

I propped my bike on its centre stand, removed the hard luggage cases, and rolled out my tools. Sarah napped while I lost myself in thoroughly cleaning each chain link before applying some much-needed lubricant.

As I quietly worked my way along the length of the heavy chain, I heard the throaty sound of a powerful motorcycle engine approaching. I looked up to see an absolute giant of a man straddling what looked like a toy motorcycle. He had spotted our bikes and deftly dodged the tents crowding the campground until he pulled alongside our tent. Then I realised he was riding a normal-sized bike and was even larger than I had first thought.

He abruptly killed his engine and began fumbling with a crutch tied along the length of his bike. Once he freed the crutch, he puffed and panted as he struggled to dismount; the whole process looked agonising. Finally, he stood, slowly unfolding to his true, staggering height, partly supporting himself with the flimsy crutch. With his other hand, he removed his helmet and, with an insane-looking grin, introduced himself in a deep, booming voice.

"Hi, my name is Zeb." He eyed the empty ground next to our tent. "Do you mind if I camp next to you?"

"Of course not," I responded, wondering if anyone had ever said "no" to this colossus.

"Very well," he replied as he began unloading his camping gear.

"Would you like some help?" I asked, watching him do everything with one hand, the other occupied with his crutch.

He guffawed at the remark. "What I look like?" he responded in his thick, Hungarian accent. "Some sort of cripple?"

He watched me carefully as I subdued a smile, and then he patted my back with his free hand, almost knocking me off my feet as we burst into laughter. After the stressful morning, it was just what I needed, and we became instant friends. Sarah popped her head out of the tent to see what all the noise was about and Zeb smiled and introduced himself. He proceeded to lighten our spirits with stories from his colourful past, explaining how he received each of the injuries that had left him with an endless list of physical complaints. Strangely, not a single story involved a motorcycle. From sky-diving to mountaineering – he had done it all – and each account of his intrepid adventures invariably ended with him lying in a puddle of his own blood with multiple broken bones.

He noticed my tool roll, lying open next to the rear tyre of my bike.

"Oh no!" he exclaimed. "You have problem?"

When I explained that I was just performing some preventative maintenance, he looked confused.

"What is this . . . pre-ven-ta-tive-main-te-nance?" he asked, carefully enunciating each syllable as though it were some dark magic spell.

"Well, I try to prevent things from breaking, before they break." It sounded ridiculous, even as I said it.

I'm sure his laughter could be heard across the entire campground and beyond. "My friend," he said, tears rolling down his face. "If something is going to break, it's going to break. Why not wait until it breaks and then fix it?"

Speechless, I blinked at him, the concept being so foreign to me.

He continued. "You should be spending your time with your beautiful lady and making the most of this wonderful day. Life is short. Believe me, I know." He tapped his damaged leg with his worn crutch as if to emphasise the point.

I smiled.

"Look at my beautiful girl." He proudly pointed at his bike. "Nothing is broken."

I glanced at his bike. It was in such terrible condition that I couldn't even identify the model or make, but the machine had sounded heavenly as he approached earlier. After Sarah's crash that morning, I strongly agreed with Zeb's sage advice. I stopped what I was doing and put my tools away.

While Zeb set up his camp, Sarah joined me sitting on a rock as I pulled out our guidebooks, and after a hug, we began to research things to do in Amsterdam.

Museums and galleries abounded throughout the city, and we had a hard time deciding which ones to visit on our limited budget. Fellow travellers had recommended two, so we set aside several days to explore and relax.

We spent half a day at the Sex Museum, on the edge of the red-light district, gazing in wonder at the endless methods, paraphernalia, art, and oddities associated with sex.

It turns out people have been doing it for years – who would have guessed?

Thoroughly enlightened, yet surprisingly un-shocked in this internet age, we left in search of a coffee shop where we could plan our next excursion. While we leisurely wandered until we found exactly what we were looking for, we ended up having an entertaining afternoon of people-watching and picture-taking.

The next day was set aside for a tour of the more refined Rijksmuseum, located on the outskirts of the city centre. Relics of Dutch history from the Enlightenment to present day, enclosed in the labyrinth of the elaborate city gateway, could have kept us enthralled for weeks. Most enchanting were the rooms filled with the works of the masters of oil and canvas – from Rembrandt to Vermeer to Van Gogh. Gazing upon these masterpieces filled us with hope, for as much aggression as the human race is capable of, there is still the ability to produce such timeless beauty. As we deciphered headlines in the local press about conflicts around the world, particularly in regions we were moving towards, the human effort wasted on war seemed all the more futile. It's been argued that without war there would be less progress, but that all depends on how we define progress.

It is a heartbreaking paradox that humans are capable of incredible innovation and unfathomable evil.

As a species, we have exhibited a great deal of ingenuity when it comes to devising ways to destroy ourselves. It is truly remarkable how much of our time, resources, and creativity have been dedicated to inventing ever-more sophisticated means of warfare. How many people, potentially even more brilliant than Einstein, da Vinci, Mozart, or Steinbeck, were never allowed to enrich our lives; their genius sacrificed upon this altar of progress; their gifts lost forever to the ravages of war, enslavement, or genocide?

The sad truth is that if we were to channel the same amount of effort towards making this world a better place for everyone, we could accomplish unimaginable things. We could end poverty, hunger, and disease while building sustainable communities that promote social justice and environmental stewardship; we could unlock new technologies that would make our lives easier, healthier, and more fulfilling. The possibilities are endless.

We must decide where we direct our energy and resources and ultimately choose what kind of world we want to live in.

***

After several reflective days in Amsterdam, we were eager to move on and leave behind the windswept lowlands of Holland. We pushed ourselves and our bikes onward and into Germany.

We crossed yet another invisible line on the earth, drawn towards the Rhine Valley, and pulled over alongside the road to Koblenz for our first night of "wild camping." We planned to do more of this as the weather improved, but to describe it as "wild" was a bit of a stretch. Pulling the bikes underneath some secluded trees near a rest stop, we were most definitely aware of the moderately busy road right beside us. As soon as we pitched our tent, we discovered that on the other side of the trees, a railroad track was constantly in use. The German nation has a reputation for punctuality, and every thirty minutes, almost to the second, a high-speed train would whizz past, close enough to buffet the fabric of our tent while shaking the ground we slept on. Aeroplanes flew overhead throughout the night, and boats blew their horns. You'd think that would have had us packing up and searching for another place to rest, but we remained.

When you are tired enough, it is possible to sleep anywhere.

After sixteen full half-hours of sleep, we rode into Koblenz, home of the Lorelei: a large slate rock that rises over a hundred metres out of the Rhine River. As the swift current caresses the base of the towering cliff, we heard a gentle murmur across the surface of the water. Legend has it, the beautifully tragic Lore Lay, betrayed by her lover, threw herself into the turbulent channel; her body lost forever. Now, her spirit sits upon the rock, combing her golden hair and distracting passing sailors with her beauty and song, luring them to a watery death.

After all the heavy rain this region had experienced of late, the Rhine was at full capacity – an impressive sight. Many of the campsites along its shores were closed due to flooding, and watching the skippers manoeuvre their barges and tour boats with grace against the relentless flow was a sight to behold. After many years of rafting all over the world, Sarah and I had developed a healthy respect for the seductive charm and deadly potential of big water.

Gradually, the weather improved on a journey which, up until now, had been beset by high winds and torrential rain. The sun emerged, blue skies appeared, and the roads began to dry. It was just in time, as our road twisted its way south, hugging the banks of the Rhine. We'd read a little about this region, and another biker had recommended the road, but nothing had prepared us for the rugged beauty of this landscape.

Bustling towns and hamlets with lively beer gardens and cafés lined the route, forcing us to ease back on the throttle and savour our surroundings. Every strategic high spot seemed to be topped by a castle, fortress, or monastery – sometimes all three at once. Some lay in ruins, but the majority appeared to have occupants still. Beside the river, at the foot of each rise, was a settlement almost in direct proportion in size to the fort above. I couldn't help but wonder if the success of the barons and bishops residing above depended on the village and its people below, or vice versa.

Sadly, the road we were so thoroughly enjoying became wider and straighter, and it was time to start thinking about where to camp. We decided to make the small city of Würzburg home for the next few days and arrived early enough in the afternoon to set up camp in a beautifully manicured campground and explore the historical Old Town.

Perched upon a hill overlooking the city was an impressive monastery that was once home to the bishop princes of Würzburg. It was a palace within a castle, within another castle, on top of other fortresses older still; each destroyed, rebuilt, and repurposed over millennia. We crossed the mediaeval footbridge of carved stone statues and cobbles stretching over the river, connecting the old town to the steep, narrow street which climbed up to the formidable castle walls. By early evening, after sampling some delicious local bratwurst, we made our way back to camp to begin planning for the following day.

As Sarah and I rode through the gates into the campsite, I noticed an older gentleman near the reservations office do a double take when he saw our loaded bikes. As we unpacked our modest supplies for that evening's meal, he approached and quietly observed us, his attention focused primarily on our motorcycles.

"Do you ride?" I asked.

"Many years before . . . and I still miss it," he responded in a heavy German accent.

I offered my hand and he shook it firmly. As the bikes cooled, the engines quietly ticking, he appeared to be struggling to find the words he needed to express himself. I sensed that he longed to get back on a motorcycle; they certainly can have that effect on a person.

"A long time ago, I travelled everywhere on my motorcycle. I loved that beautiful machine and the freedom it gave me," he finally said.

"Why did you stop?" I asked.

"Oh, you know. Life, responsibilities, children." He shrugged, walking a slow circle around our bikes.

Then he paused, pointed at himself, and said, "My name is Hans, and I own this campsite. I get to meet lots of adventurous people like you, and it allows me to relive my youth when I hear about their travels. Can you tell me about your journey?"

"Please join us," Sarah said as she cleared a space at the picnic table.

He sat with us as we prepared dinner, and we shared a bottle of wine, listening to him reminisce about his adventures as a young man on his motorcycle before he decided to settle down. He seemed fascinated with our plans and was particularly impressed that Sarah was on her own bike. He remarked that most couples rode two-up.

I told him about our journey through Central and South America on one machine and Sarah's stubborn determination to ride her own for this journey. As an accomplished river guide, Sarah was a natural leader. On a single motorcycle, there can be only one person in charge of the necessary split-second decisions. I was aware of Sarah's frustration multiple times during that trip that she wasn't in control. Soon after we returned from that adventure, I found Sarah, her finger on her bottom lip, thinking as she walked around the motorcycle. She noticed me and looked up.

"Teach me," she said, gesturing towards the bike.

"To ride?" I asked.

She nodded, and climbed onto the front of the bike, implying that I should ride pillion.

My first attempt to teach her how to ride didn't go as planned. We found a deserted car park, and I ran her through the basics. She had picked up a surprising amount of knowledge from riding as a passenger, but the weight of the bike was initially too much for her. When I finally encouraged her to take full control, she smoothly accelerated, performed multiple turns, and manoeuvred through an improvised obstacle course we had set up before coming to a controlled, deliberate stop. Without momentum, though, she lost her balance, and the bike crashed to the ground, narrowly missing her leg.

Watching a person you deeply care for almost hurt themselves can be enormously stressful.

The fall had bruised her confidence, and she was reluctant to ride again. We agreed that a professional instructor and a lighter bike would probably be wiser for her – and our relationship. We found exactly what we needed on our return to Colorado: a weekend-long course that taught some theory and a lot of practical skills, all on lighter, lower machines. After the intensive program, I was shocked by how much she had improved. She would amaze me every day with how confidently she handled her bike. She seemed to be a natural.

Besides, it made a lot of sense for Sarah to ride her own bike, and we should have addressed it before setting out on that first motorcycle trip together. If I had been injured or otherwise incapacitated, she would have been, effectively, stranded.

We weren't going to take such a risk again.

Hans sat transfixed as we related our story, and then, as the sun set and the temperature dropped, he rose from his seat and wished us a peaceful evening. When he turned to leave, he hesitated as something he remembered triggered a huge grin to spread across his weathered face.

His eyes sparkling mischievously, he added, "While you are in this region, there is one road you really must take."

It turned out to be one of the best tips we had received.

After storing all of our unnecessary gear out of sight inside our tents, we let our now lighter and more agile bikes impress us yet again with

36

their full potential as we followed Hans's advice. The B22 was a lonely, two-lane, well-surfaced road winding through lush countryside dotted with vineyards and sporadic sleepy villages. Occasionally, too awestruck to zoom by, we slowed our bikes to a crawl to marvel at the magical scenery; from time to time, whoever was in front would glance back at the other in a "Are you seeing this?" sort of way.

After two hours of glorious riding, we arrived in Bamberg, smiling from ear to ear.

Bamberg turned out to be just as impressive – and possibly more – than Würzburg, with architectural relics on every street. The city was working hard to maintain and restore what remained of its heritage, and I found it hard to imagine what building codes would apply to structures as ancient as these, with walls up to a metre thick. Our only regret in Bamberg was not allowing ourselves a taste of their unique, local smoked beer. With a long ride back to camp ahead of us, it was a pleasure we intended to enjoy on another day.

***

One of the lessons we were slowly learning was that effective communication would be vital for us to succeed, both with the people we met along the way and with each other.

With each day that passed, we discovered more about ourselves and our relationship.

At first we assumed we were on the same page about our needs and wants, but our expectations were often out of sync. With a great deal of intentional effort, we were beginning to understand each other better and appreciate our strengths and weaknesses. We would often chuckle when we recalled one of our silly disagreements over describing a shower token used by a campground earlier in our trip.

"Can you pass me a round token, please?" I had asked as I stood lathered in soap, my eyes closed beneath the stinging suds. I was in a campground shower stall, and the water had suddenly stopped just as I was beginning to rinse myself. Sarah was in an adjacent stall and held the tokens I needed to restart the water.

"What round token?" came the response. "I don't have any round tokens."

"Yes, you do!" I answered sharply.

I now had soap in my eyes, ears, and mouth, and I thought she was deliberately testing my patience. We had just purchased enough tokens for showers and laundry, and I couldn't believe she had already used them all. I sighed exasperatedly and exclaimed again, "The round tokens for the shower. You have them all. Please, Sarah, can I just have one? I have soap everywhere."

"I don't have any round tokens!" I could hear from her voice that she was now losing patience with me. "The tokens I have are square."

"What? Round, square, whatever, can I just *have* one?!" I pleaded. I couldn't imagine why she was being so difficult. Suddenly, I heard my shower curtain open abruptly, and I turned to face the sound, squinting through soapy eyelids at whoever had just intruded. It was Sarah, and she was holding out a token.

"Here!" she cried, thrusting a token into my hand. I couldn't decipher its shape as I fed it into the coin box, and instantly, ice-cold water drenched me.

"Holy *shit!*" I gasped as the freezing deluge took my breath away.

After we had showered and given our tempers a moment to wane, we quietly prepared our evening meal. It all seemed so trivial once we had food in our stomachs and were warm and dry. We realised we had been on the road together for over a month, constantly in each other's company. The stress of the rain, cold, long days spent driving, and lack of adequate sleep amplified our irritating habits and idiosyncrasies.

Sarah produced a shower token and placed it on the table between us. "Tell me what exactly you think is round about that," she said, nodding at the token.

The token had been round originally, and then two opposite edges had been trimmed. Was it a circle with two square edges or a square with two round edges? If anything, it looked more like an oval. We were able to laugh about the misunderstanding afterwards, but it definitely taught us a life lesson. It was a subtle reminder of how differently two people can view what is around them and how easily that can lead to confusion and disagreements. We promised each other to practise more patience and tolerance.

Holding hands across the picnic table, we gazed at each other with so much compassion and understanding. I never wanted to be angry with my beautiful partner of ten years. I loved being loved by her. She was tough, and there was a gentleness to her that I aspired to have. She was a leader and she trusted me to lead the way. She was smart and responsible and had given up her career, financial security, and practically everything for adventure – just like me.

As I was thinking about what I loved most about my wife, she squeezed my hand and smiled at me, the corners of her warm hazel eyes crinkling. I cupped her face with my spare hand, my gaze wandering from her smile lines to her slim, delicate nose with the faintest hint of freckles from time in the sun. I caressed her soft, perfect skin, lightly tanned from constantly being outdoors.

There was so much I loved about her personality. And it definitely helped that she was stunningly beautiful, too.

***

After leaving Bamberg, we again followed the advice of Hans and took the scenic route to Heidelberg. We were thrilled to drive a wonderful, twisting, empty road that led us into the novel little city on the banks of the Neckar River. We arrived late on Sunday, with just enough time to explore as darkness fell. The cool evening provided a welcome respite from one of the hottest days we'd had to date.

By now, the bikes were overdue for some attention, with the oil needing changing, filters begging to be cleaned, and chains itching for lubrication. We decided to establish ourselves for a few days outside a major city where we could find the necessary supplies. Stuttgart was nearby, so we wound our way up into the Black Forest and stumbled on the charming little village of Höfen. We found a comfortable campsite on the banks of a crystal clear creek. We tracked down a BMW motorcycle store in the city, and as always, felt thoroughly violated after paying a premium for some basic supplies. But back at our campsite, we were able to take care of all our basic bike maintenance needs in an environment we appreciated. Besides, I always found working on our bikes therapeutic.

Nothing could ruin our mood that day.

Even when a magnificent thunderstorm broke loose overhead during the oil change, I shook my head and smiled at Sarah as I wrapped up my tools. We dashed for cover from deafening thunder as bolts of lightning illuminated the grey sky amid a torrent of rain.

Laughing as we changed out of our wet clothes inside our tent, we collapsed onto our mess of sleeping bags and gear and waited out the storm wrapped in each other's arms. Although it was hard to hear over the noise of the downpour pummelling the thin walls of our lightweight shelter, we reminisced on our favourite parts of the journey so far and shared our excitement for what was ahead.

"Pinch me," said Sarah.

"No, *you* pinch *me*. I still can't believe we're doing this," I replied.

"Every day," Sarah mused, "when I hop on that bike, I think of how lucky we are."

"Every day, when you hop on that bike, I think of nothing else but how good you look in your riding outfit," I said, with Sarah's soft laughter punctuating my words. "Especially with motorcycle grease on your nose."

Sarah punched my shoulder lightly and rubbed her nose.

At that moment, we noticed the storm had stopped. We climbed out of the tent and resumed our motorcycle maintenance tasks. It was always satisfying to complete a full service, and both machines responded well to the attention.

We fondly recalled Zeb's philosophy: "Don't waste time on preventative measures. Wait until it breaks, then fix it." We considered applying this logic to one of our bikes and comparing the results, but we couldn't agree on whose bike we wanted to neglect. It seemed strange that, even though they were just machines, after riding them for the best part of each day, it was difficult not to become somewhat attached to them.

I reflected on how I sometimes allowed myself to practise this laissez-faire approach when it came to life and my relationship with Sarah – burying my head in the sand and waiting for things to break before trying to fix them. I didn't want to take what I had for granted, like Zeb and the art of motorcycle maintenance, that left him injured and alone.

CHAPTER FIVE

Room for Improvement

Personal Diary Entry: 3rd July 2013
Day 51 (of 549)
6,943 km of 62,840 km (4,314 mi of 39,047 mi)

As we continue on our odyssey, the experiences we encounter are both diverse and enlightening. Every day brings new faces, places, and challenges that we approach as a team. We've shared laughter and tears, moments of triumph and defeat, but through it all, we are growing closer to each other. It is as if this journey is a mirror, reflecting back to us the best and worst parts of ourselves, while forcing us to confront and overcome our own limitations.

The city of Stuttgart turned out to be a little disappointing aesthetically compared to where we had been over the previous few weeks. It felt a little too officious and business-like, more impersonal. For the first time in Germany, we began to notice extremes of wealth and poverty, the destitute begging for change outside boutique stores selling overpriced trinkets to the über rich. After a few more days in the region, we broke camp and moved further east.

Augsburg became our base for the next five days. A sizable city close to Munich, we felt it would be an excellent location for getting a feel for the Bavarian side of Germany.

A day trip into Munich coincided with a BMW-sponsored marathon, live music, and a miniature beer festival. A party atmosphere filled the streets, and we couldn't resist indulging in some of the local beverages. Bands played traditional music as muscled young men in tight *lederhosen* cracked leather whips to the beat. Sarah seemed quite impressed – squeezing my hand with a look of sheer delight.

We followed our visit to Munich with a sombre day at Dachau, immersing ourselves in the remains of the first concentration camp from the Third Reich era, a prototype for the horrors that followed. Shuffling through the compound in stunned silence, we were overwhelmed by what we were experiencing, struggling to control our emotions. Often, we had to stop and sit down and breathe.

It was hard to imagine how such a place came to be.

A chilling museum within the camp reflected on the power of propaganda on a population whose economy lay in ruins. Martin Niemöller's poem, "First They Came," hung over the faded pictures of those who suffered under conditions of sickening barbarity. Gaunt faces with eyes devoid of hope stared at us from the past, pleading never to allow this nightmare to happen again.

We had been hesitant about visiting this place. I wasn't comfortable with its status as just another thing to do while in this area. Its significance was so much more important than that, and the horrendous power of what it represented affected us deeply.

We didn't talk to each other for the rest of that day.

What was there to say? I think we needed time to process what we had experienced. As we held each other closer that night, I could feel Sarah's warm tears on my shoulder as she quietly wept.

***

As always, with motorcycle travel, things wear out and need replacing; and not just on the bikes. Our next project involved chasing down a new set of tyres for both bikes, four in all. Stricter regulations in Germany prevented us from buying the specific models we wanted there. Luckily, with the help of a friend in Switzerland, we were able to order the tyres we preferred and have them shipped to his place. where we planned to have them fitted on our arrival.

We had become intrigued by the remnants of Europe's rich and elaborate history, and we found it nearly impossible not to pull over at each one we encountered.

While riding over the Ammersattel Pass and through Linderhof, we stopped to investigate one of King Ludwig II's extravagant palaces. The gardens surrounding this particular palace, still considered one of Europe's most beautiful creations, attracted most of our attention. We spent a tranquil morning strolling through the manicured estate, pretending to be esteemed members of some long-forgotten royal family while dressed in our scruffy riding attire. Adorned in dust from the road and with more than a hint of dishevelment, we sauntered through the grounds, giggling at our ridiculous role-playing. With exaggerated airs and theatrical gestures, we whispered grandiose proclamations as though commanding an imaginary court. I wondered if the original residents would have applauded our audacity or summoned the palace guards to have us thrown out.

I envisioned the latter, with us flat on our arses outside the palace grounds.

As we returned to the car park, we bumped into a group of local riders gathered around our machines. We talked for some time about our shared passion and the call of the open road. They offered insights on their favourite rides in the region before alerting us to the upcoming BMW Motorrad Festival in Garmisch-Partenkirchen. We had a rough plan of where we'd like to be by then – somewhere in Switzerland – a good deal further along.

But we decided to break our own rules.

Before setting off for this trip, we agreed that we would travel with an open agenda so we could follow the advice and suggestions of locals we encountered along the way. So we made a plan to pick up our new tyres in Switzerland and then return to Germany later for the festival.

While in the region, we couldn't resist taking a short detour to visit the fairytale castle of Neuschwanstein, the inspiration for the

45

iconic Disney logo. Commissioned by the reclusive Ludwig II to be used as a personal refuge, it dominated the landscape, overlooking the beautiful village of Hohenschwangau. It allegedly cost him his entire personal fortune – and possibly his sanity. Sadly, he didn't live to see its completion.

After securing our bikes in the car park, we came upon one of the longest lines we'd ever seen with an astronomical wait time, just like the rides at Disneyland. We decided the guided tour wasn't for us. Instead, we spent a few moments gazing up at the formidable landmark nestled into the foothills of the Alps, imagining tales of pale princesses and noble knights. Then, we retreated and spent the day hiking through the surrounding mountains, taking a much-needed break from our long days in the saddle.

Besides, the castle looked even more impressive from a distance.

Each time we camped in a new region with a day off the bikes, we squeezed in some activity to stay physically and mentally healthy. Riding a motorcycle takes a certain kind of fitness. It can be quite demanding, especially on the more challenging roads. Unfortunately, it wasn't giving us the cardiovascular workout we needed, so strenuous hiking became our on-foot exercise of choice.

Suitable trails were everywhere, and locals were more than happy to point us towards their favourites. We quickly became accustomed to the charming huts that appeared at the top of many of the highest peaks, each one a safe haven from the changeable weather and a great place to pick up a cold beverage on a hot day.

From Germany, we followed more local advice and rode into the clouds enveloping the Hahntennjoch Pass. The higher we went, the more we were bathed in the mist that kept the roads a slick and shiny black. At the summit, our feeling of exposure was from more than the freezing temperatures; we knew that precipitous drop-offs were hidden in the clouds around us. As we took in the fantastical scenery, grey fog replaced the white mist, and it began to rain.

It was magical, but soon we were soggy and shivering.

We rode on and dropped down into Austria. There, we took a short detour towards Innsbruck, seeking warmth, shelter, and Wi-Fi. We agreed to pull over at the first coffee shop we encountered – which turned out to be a Starbucks. Reluctantly, we entered and ordered our hot drinks, happy to be out of the chilling rain.

"Feeling a bit pretentious today, are we?" said Sarah dryly, when I brought our overpriced coffees to the table she saved for us.

I nearly choked on my decaf-soy-milk-low-fat-orange-mochaccino.

By late afternoon, we entered the town of Scuol, Switzerland. After many kilometres on the road, and especially in miserable conditions, seeing an old friend is soothing to the soul. Whitewater legend Kyle Spinney was expecting us in his home on the banks of the Inn River sometime soon. However, as our estimated time of arrival didn't fall on a specific date, when we rode in, he was entertaining clients on a rafting adventure.

But there was no need to fret.

Kyle's partner-in-crime, Hamish, greeted us with a hospitality so warm and magnanimous, that we soon forgot all about our discomfort. He opened his doors and, within seconds, made us feel like long-lost friends. Later, when Kyle had finished his rafting trip, he returned home, and we had an emotional reunion, possibly influenced by Hamish's exceptional selection of fine whiskeys. We sipped single malts until after midnight, when, still in our motorcycle gear, we finally called it a day.

Kyle's home was where we'd had our new tyres shipped. Two of our tyres had arrived, but the others were still in transit, so we spent the next few days rafting and hiking, and enjoying the spectacular village of Scuol. Voted one of the most beautiful towns in Switzerland, it was originally established as a health spa. On every street corner, spring water flowed from ancient fountains, each with unique properties. Some were even naturally carbonated! We gorged ourselves to rehydrate after weeks on the road. Thankfully, Kyle had advised us to use caution when drinking from one of the fountains, renowned for its unique cleansing properties. It was situated suspiciously close to a busy public toilet.

Eventually our tyres arrived, and after having them fitted, it was time to hit the road again. My dear friend Kyle and his housemates provided a list of recommendations for roads and passes to take us back into Germany. We stitched together a route encompassing seven must-see Alpine passes, looping down into Italy, back through Austria, and north into Bavaria.

We were eager to get back on the road.

It took us two days of magnificent riding to reach Bavaria, but by the end of the first day, our luck with the weather began to run out again. It all happened so suddenly that we had no time to plan ahead – the skies simply darkened, the heavens opened, and rain thoroughly soaked us within minutes, finding every exploitable crack in our weatherproof armour. As we slowed our pace because of the worsening visibility, I glimpsed what looked like a campsite behind a heavily wooded area next to the road.

One of the luxuries we had invested in for this journey was a set of helmet intercoms, allowing us to easily talk to each other while on the road or listen to music. Sarah liked listening to music rather than my voice, so I had the intercom off for most of the ride.

I tapped the side of my helmet to turn on the two-way radio. "Did you see that?!" I yelled over the sound of the rain.

"See what?" she replied. "And stop *yelling* at me."

"I think that was a campsite we just passed on the right. Shall we turn around to get out of this rain?" I said, hoping it hadn't just been wishful thinking.

"Sure, let's check it out," came the response.

We pulled over and checked for oncoming traffic before performing a sweeping turn back in the direction we came from. I spotted an entranceway but no signage indicating where it led. As we exited the main road onto a forest track spongy with pine needles, I began to doubt my decision.

"Are you sure you're right about this?" Sarah asked, a hint of frustration in her voice.

"Aren't I always?" I snapped back sarcastically.

After a long day of hard riding, we both felt quite drained.

I'd been wrong so many times during this trip so far – when it came to directions, weather, or how much sleep we could get when camping next to a railroad – that I could hear Sarah laugh through the intercom.

A moment later, the woods opened to reveal what looked like a long-abandoned campground with a central timber structure surrounded by recessed tent sites. We came to a stop outside the main building, turned off our engines, and took shelter under the awning by the front door. Curtains of rainwater poured over gutters clogged with detritus. We peered through window panes caked with dust and cobwebs. Nothing moved inside the unlit interior of the building.

"Hello!" Sarah called as she knocked on the door.

There was no response. She reflexively tried the handle, and the door swung open. We looked at each other and shrugged our shoulders.

"Helloooo!" she called again, this time into the quiet room within.

We stepped inside, thankful to be out of the horrendous rainstorm. The light was fading now, and a chill was setting in. The prospect of getting back on the road held little appeal. With its door unlocked and its dry interior, this building felt like it had been placed here just for us to take a rest in. Even though it seemed to invite us in without needing a camp host, I was nervous that we were trespassing on private property. I couldn't understand why a public campground would be closed in the middle of summer. I expressed my concerns to Sarah.

"Well, why don't we stay here for a little while, maybe dry off a bit, cook some warm food, and see if anyone turns up? I'm sure they would understand," Sarah suggested.

"I'm not sure if we should. I don't want to get us in trouble," I replied, scanning the forest for any sign of the owners.

"Oh, come on, don't be so paranoid. You're the one always telling me I should be a bit more reckless sometimes. Where's the harm? Let's be reckless!" she said, giving me a seductive wink.

A thunderous crack of lightning was all it took to shatter any reservations I had about bending a few rules. We quickly unpacked our gear and brought in a change of clothing and what we needed to prepare our evening meal. Before long, we were enjoying a simple yet deliciously hot meal, cosy inside our fresh, dry clothes.

"Do you know what day it is?" Sarah asked in between mouthfuls of pasta.

"July third," I offered.

"Hmm!" was the response I received, leaving me wondering what I had missed.

It hit me a little too late.

On the sun-drenched morning of July 4th, exactly ten years after Sarah and I had first met in a small bar in California, we pulled into the festival we'd come back to Germany for. Suddenly, her question from the night before, tinged with a subtle expectation, unfolded its layers. We'd always considered that date to be an anniversary of sorts, and here I had missed yet another one.

As diligent as I was with preventative maintenance on the bikes, I had room for improvement in our relationship. I knew I was prone to slipping up occasionally, and this moment highlighted my inadequacies as a supportive and considerate partner. Choosing our route and destination each day was relatively simple compared with navigating the labyrinth of human connection, which demanded a finesse I had yet to master.

I vowed to do better.

Holding Sarah tightly as I reassured her, I felt some tension ooze out of us as we reconnected and recalibrated in preparation for an eventful few days.

CHAPTER SIX

Maths and Beer Don't Mix

Personal Diary Entry: 13th July 2013

Day 61 (of 549)

7,620 km of 62,840 km (4,735 mi of 39,047 mi)

As the kilometres accumulate behind us, a vast expanse of unknown territory still lies ahead, shrouded in anticipation and uncertainty.

In two days, we had ridden some of the most acclaimed motorcycle roads in Europe and arrived at the biggest BMW owners' gathering we'd ever seen.

From Thursday, July 4th, to Sunday, July 7th, the German city of Garmisch-Partenkirchen in the Bavarian Alps played host to the annual BMW Motorrad Days Festival. That year, BMW celebrated its 90th year in the motorcycle production industry with an event to remember. With an estimated 35,000 riders and enthusiasts from all over the world in attendance, the streets, campgrounds, and hotels were swarming with every BMW model, old and new.

The event site, located a mere eighty kilometres from Munich, covered an expansive area, becoming a village of vendors, attractions, exhibitions, and beer and food stalls. A further 15,000 square metres of fully serviced campgrounds surrounded the site, creating a suburb of nylon and canvas. The event organisers had drawn on the expertise acquired over decades from Munich's Oktoberfest celebrations, and there was enough food and beer to satisfy an army of adventure riders.

A large contingent of bikers from Malaysia completed their overland tour by riding into the event on Friday afternoon. We saw licence plates from many different countries, but only two others were from the USA.

We recognised the bikes immediately.

They belonged to Dawn and Paul, who we had camped beside in Belgium, and we patiently waited by their machines, hoping for an impromptu reunion. We had no idea they were planning to attend the same event, and we bubbled with excitement at the prospect of seeing their familiar faces again.

"Sarah . . . Irish!" Paul's familiar voice called out as they approached us through the bustling campground. We shook hands and hugged, congratulating each other on making this far. It seemed oddly natural to have formed such a strong connection with two people we barely knew. Even though it hadn't been that long since we had seen them, the odds against running into them again in the middle of such a huge event seemed staggering. It felt good to have companions as we immersed ourselves in the exhilarating weekend ahead.

Months before, at a Horizons Unlimited gathering in California, I had convinced the editor of an adventure motorcycle magazine to issue press passes to Sarah and me. I had hoped, at the time, that they could help open doors or grease a few wheels while we were on the road.

Now was our chance to put the press passes to the test.

When I presented our credentials to the crew on duty at the press tent, they gave us VIP wristbands, allowing us access to all the exclusive behind-the-scenes areas. With those wristbands came a multitude of privileges, including complimentary food from the onsite vendors and, of course, the coveted gift of free beer.

For approximately half a day, I took my responsibilities seriously, making fastidious notes about everything we saw and experienced. Then, temptation beckoned, and the allure of the complimentary beer grew irresistible. I succumbed to the indulgence. As the amber elixir flowed, my note-taking gradually lost its meticulous organisation, giving way to a more relaxed approach. But writing down gibberish was the least of my worries.

Maths and beer don't mix well, either.

My fail-safe idea for determining the exact number of motorcycles present – counting all the wheels and dividing the total number by two – was not as foolproof as I had thought, given the presence of a considerable number of three-wheelers.

Ambling through the event grounds, I marvelled at all the shiny new toys on display. The air was charged with excitement as I revelled in the spectacle. Meanwhile, Sarah, with her keen eye, scanned the crowds for more handsome young men with whips and tight lederhosen. She seemed to have developed a newfound interest in traditional Bavarian music.

A purpose-built off-road enduro park allowed riders to test a selection of bikes from the GS fleet, from the mighty 1200cc Adventure to the nimble 650cc Dual-Sport. Under the expert guidance of the BMW instructors, riders with varying degrees of ability could experience what these bikes were capable of in more challenging terrain. We put our skills to the test as we manoeuvred the bikes through muddy puddles, over rugged bumps, and across untamed landscapes. The symphony of growling motors, the adrenaline coursing through our veins, and the thrill of conquering nature's obstacles united to create an unforgettable experience. We had so much fun we couldn't help but attempt the course multiple times.

For those wanting to experience how the motorcycles handled on the street, the entire current fleet of BMW road bikes was made available for test rides, and for more than a quick spin around a car park. After so many weeks on our fully-loaded enduro bikes, we couldn't resist riding something lighter and more responsive. The F 800 R, the K 1300 R, and the spectacular 193-horsepower S 1000 RR provided all the excitement we were looking for, and more.

Our spirits soared atop each magnificent machine.

The powerful engines beneath us responded eagerly to every twist of the throttle, propelling us on with unbridled energy at what felt like the speed of light. Postcard-perfect scenery flashed by as we weaved our way with precision, leaning into each turn, fully immersed in the moment.

I felt like I had broken the space-time continuum. Somehow, I was able to savour the harmonious blend of adrenaline-fuelled racing and the tranquillity of my natural surroundings.

Back at camp, various shows and displays entertained the crowds. Gravity-defying acrobats sped over the creaking wooden planks of the world's oldest Motordrome. Stunt legend, Chris Pfeiffer, performed tricks in the event arena, even putting the cumbersome 1200 GS through a seemingly impossible sequence of manoeuvres. An onsite cinema featured inspiring movies of motorcycle journeys from around the world, and each evening culminated in an enormous party with live bands, motorcycle raffles, and a staggering amount of food and beer. For those who preferred a more relaxed evening, talented musicians provided live guitar music around the open campfires.

On Sunday morning, as the Metropolitan Jazz Band from Prague entertained the breakfast crowds, the weekend event gradually began to wind down. Later that afternoon, swarms of motorcyclists took to the roads in every possible direction; some beginning, some ending, or, like us, continuing their adventures on two wheels.

We reluctantly packed up on Monday morning and, based on recommendations from several bikers we'd met over the weekend, took an exciting series of Alpine passes to Innsbruck in Austria. That evening, what we'd thought would be a quiet night at a small campsite turned out to be an extension of the party atmosphere we'd left behind in Germany as several bikers joined us from Israel and the UK. A shaky start the following morning saw us accepting yet more advice and climbing over even more spectacular winding passes back into Switzerland to our friends in the village of Scuol alongside the Inn River.

As welcome as they made us feel, we sadly had to push on as our rough schedule for Europe was beginning to get pushed further behind. So after one final night in Scuol, and a few too many glasses of exquisite single malt shared with our friends, Sarah and I prepared our bikes for an early departure.

On waking, we fired up our engines and took to the road.

CHAPTER SEVEN

Like Living in a Postcard

Personal Diary Entry: 30th July 2013
Day 78 (of 549)
10,820 km of 62,840 km (6723 mi of 39,047 mi)

The storms have tested our resilience, as well as our ability to adapt and work together as a couple. We are learning the value of patience and perseverance, and the importance of taking time to rest and recover when needed.

We know the challenges we face now are only the beginning, and that the road ahead will be filled with even greater obstacles. But we are ready and willing to face them together, fortified by the experience we have gained and the bond we have formed through our shared trials.

The road to Interlaken, from Scuol, took us through yet more spectacularly deep valleys and over some impossibly high passes. Locals had advised us it would take approximately five hours, and it took us two full days because of numerous stops. Each twist and turn revealed a panorama that made it hard to keep our eyes on the road, so we often stopped to gaze upon our surroundings, snapping pictures and snacking on bread and cheese.

The changing weather conditions would tease us with glimpses of the snow-capped peaks that flanked our route. By the end of the first day, we rode into camp thoroughly exhausted, having covered a little over half the distance we'd intended to. A hearty meal, followed by a good night's rest, set us up well for the following day's ride, and by three in the afternoon, we were coasting alongside the deep turquoise waters of Lake Brienz on the final stretch into Interlaken.

This mountain town had become the home of Switzerland's adventure tourism industry and a magnet for whitewater rafting guides from all over the world, many of whom we had worked with over our years spent "pushing rubber." We set up camp, knowing we'd be there for a few days, excited at the prospect of catching up with some old friends.

Faces I hadn't seen in almost fifteen years made our stay a special one.

We spent evenings at the local Irish pub, raising glasses to guides no longer with us and briefly contemplating the risks of being a thrill-seeker. The solemn mood was short-lived, however. Soon, we had our glasses refilled and were swapping stories, the way an evening is typically spent in an Irish pub.

I was often drawn towards Irish pubs when in new surroundings, as most cities and towns around the world would have at least one. When done well, I could immerse myself in an entertaining evening reminiscent of home, but a few left much to be desired. I recalled walking into one many years before in New Zealand where a sign on the door caught my attention. "No work shirts!" it demanded in bold letters. The pub was quiet inside, and the bright lights emphasised its emptiness.

"So what makes this an Irish pub?" I asked the barman.

"Well," he said, "we have Guinness on tap." His eyes swept across the room. "And an Irish flag," he added, pointing to the opposite wall, "over there."

"Then what's with the sign on the door?" I probed. "The one saying 'No work shirts!'"

He looked confused. "Well, we don't want any ruffians in here."

"Have you ever been to an Irish pub?" I shook my head at him. "We often go to the pub after work, sometimes during work, and occasionally before work. In our work shirts."

In that moment I saw a new career opportunity appear before me: the official "Irish Pub Inspector." I would spend the rest of my life visiting all the Irish pubs in the world, sampling their food, whiskey, and beer – for free, of course. For a modest fee, each one would receive a certificate of authenticity after my visit, five shamrocks for the real thing, and one shamrock for any that didn't make the cut. Perhaps I could get Jameson and Guinness to sponsor the endeavour.

Not a terrible retirement plan.

***

Arriving in Interlaken on a weekend when bookings were high and guides were scarce was in our favour. I was put to work on two of the local rivers, sometimes guiding and, other times, safety kayaking. On whitewater rafting excursions, a safety kayaker paddles ahead of the raft trip as an extra set of eyes to probe unfamiliar rapids and make available a faster rescue platform. The added income from the work I picked up helped balance our tight budget in this expensive country. Meanwhile, Sarah went canyoning in the Chli Schliere Gorge.

While in the area, we couldn't resist riding more of the local roads, and we soon climbed up towards the town of Grindelwald, which provided an impressive view of the infamous North Face of the Eiger.

Nature unveiled its grandeur in full splendour that day.

A deep blue sky contrasted sharply with the shimmering white peaks and the dark rock walls towering over the picturesque mountain town. Like majestic sentinels, the mountains reached for the heavens, their pristine beauty demanding our attention. We were mere observers of a grand spectacle unfolding for aeons, humbling us with the vastness of the world and our fleeting place within it. We had spread our jackets out beneath us as picnic blankets and sat in awestruck silence as we attempted to fully appreciate the moment.

"Can you *believe* this?" Sarah finally whispered, gesturing with a sweep of the arm at the scene before us. She seemed lost for words, as did I.

"Do you think this could be the place?" I asked.

Before we began this journey, we agreed that if we found a place we fell in love with, we could just stop. It didn't have to be for good, but perhaps, for a while, we could make it our home.

"It ticks a lot of the boxes for me," she said in a dreamy voice. "Would it be wrong to end our trip so soon?"

"It's certainly something we could consider." I put my arm around her and gave her a squeeze.

We quietly pondered the idea as we prepared a simple lunch of tasty local cheese on thin slices of apple, all wrapped inside a freshly baked, crusty baguette. Although it was a plain meal, it had become a staple for us, and we thoroughly enjoyed it every time.

"We should keep moving," I heard her say after she'd finished chewing her last bite. I turned to look at her, and she flashed me a reassuring smile, a few small crumbs clinging to her delicate lips. She paused for a moment, then continued. "There is still so much to see, and it just keeps getting better and better."

Reluctantly, we tore ourselves away to explore further. We would have been disappointed to end the trip so soon after it had begun when there were so many adventures ahead of us. So, we filed Switzerland away as a potential future home.

Mountain fever had set in.

We simply couldn't get enough, so Zermatt, a resort town at the foot of the pyramid-shaped Matterhorn, became our next destination. The direct route via the drive-on train through the mountains didn't appeal to us, so we decided to skirt around the foothills, making a slight detour into the city of Montreux to catch one night of its famous two-week-long jazz festival.

An old friend, Biff, and his son, Yarin, decided to join us for part of the ride, so we convoyed over the Saanenmöser Pass and down into the Rhone Valley. We hugged the shores of Lake Leman before riding into a carnival of live music, street performances, and every kind of food we could dream of. Yet another festival coincided with our journey – how nice of the locals of Munich, Garmisch, and now Montreux to arrange things so.

We couldn't wait to see what the rest of the world had to offer.

After a day and night of fine music, great food, and sweltering heat, we packed up and left Montreux to its revelry, determined to reach higher elevations and cooler weather. After all the time we'd recently spent shivering, it was surprising how we missed the cold already after only a few sunny days.

As we rode further into the French part of Switzerland, we noticed the landscape subtly becoming hotter and drier, with vineyards clinging to the slopes surrounding idyllic mountain villages. Unfortunately, the closing of the road at the small town of Täsch thwarted our plans to reach Zermatt. To proceed, special permission was required from the local police, forcing those wanting to visit Zermatt to take the overpriced train or hire a taxi service which used the perfectly adequate access road. Only those wealthy enough to own property in Zermatt could use their personal vehicles to proceed, so the worn-out taxi vans shared the road with a dazzling display of Bentleys, Bugattis, and Ferraris.

We briefly entertained ignoring the "no private vehicles without a permit" sign and illegally using the road. I wondered how much trouble we could really get into on our motorcycles, but the Swiss are sticklers for rules, so we reluctantly took the train. Once in Zermatt, we observed that the streets were so narrow that the town centre could only accommodate miniature, battery-powered shuttle vehicles, giving it the feel of a theme park.

But this wasn't Disneyland.

Unlike the twenty-four-metre-high Matterhorn Bobsleds, the first rollercoaster-style attraction at California's Disneyland Park, the real Matterhorn stood 4,478 metres tall.[2] The rollercoaster was modelled after the pyramid-like shape of "Europe's most famous mountain."[3] With the prospect of hiking a part of the mountain impossible to resist, we established ourselves at a nearby campground for three days, determined to make the most of any break in the weather.

On our first hike, although it was hot and humid, a heavy layer of cloud denied us a view of the crooked pinnacle that towered over the village, concealing the true scale and majesty of the mountains surrounding us. Occasionally, we glimpsed a hint of a bluish glacier catching the sunlight at the end of an adjoining valley or a jagged peak thrusting through the cloud base.

We hiked regardless of the conditions, enjoying the crisp air and the intermittent showers. Flowers were in full bloom, and the steep trails from the valley floor soon had us puffing and panting, yet thankful for the exercise. Each evening, an ominous bank of dark clouds slowly crept down the valley. Distant claps of thunder preceded the first fat drops of heavy rain, and we huddled in our cosy tent, sheltering from the impending storm, anticipating the lightning display that never failed to impress.

Only on the third day did the weather finally clear, and we were rewarded with stunning, early morning views of the Matterhorn, with its surrounding peaks, and the enormous glaciers that blanketed their bases. We chose a route that led us along the northern edge of this postcard peak, craning our necks to fully appreciate its scale and beauty.

By late morning, the clouds began to return, and as we neared the base, the summit became hidden behind columns of vapour. From time to time, the veil of moisture would slip seductively from the mountain's shoulder, revealing a possible route to the summit. We gazed in awe for a long time, fully enthralled by the power of its attraction.

It took a lot of willpower to resist climbing further.

Switzerland had effortlessly swallowed more of our time and budget than we'd anticipated, so reluctantly, we decided to move on. People have often said that living in Switzerland is like living in a postcard, but postcards are flat and this country was anything but. Except when it came to humour, which could certainly be a little two-dimensional.

And then there were the languages – this overachieving country could not suffice with only one. Like a linguistic smorgasbord, one could order their daily dose of hellos and goodbyes in a variety of flavours. Yet Switzerland knows how to make things work with breathtaking efficiency; there is a reason this country is renowned for its timepieces. It's like a well-oiled machine where one can only marvel at the seamless operation of Swiss life.

***

Leaving the Alps was difficult. We had become accustomed to powering our bikes into the endless twists and turns, utilising every square inch of tyre tread, exhilarated by the sheer exposure and breathtaking scenery. As we rode west, the terrain flattened, and the ride became more about reaching a destination than enjoying what was in between the start and finish. Alpine forests gave way to open farmland, and the horizon stretched further beyond us with each kilometre. We based ourselves on the outskirts of Paris, hoping to visit the city and use our time there to fix our broken laptop, which was essential for diary-keeping.

The computer repair shop employee told us we would have to stay a few extra days while he replaced our keyboard, so we decided to make the most of it and explore Paris. For most of the journey, we avoided large cities, but neither Sarah nor I could resist the urge to be tourists for a day or two. For our first excursion of the journey on one motorcycle, we headed into the heart of Paris early in the afternoon, and it began as a wonderful experience.

On a single motorcycle, we were practically invisible.

In most European cities, motorcycle riders can get away with a lot without being seen. Other drivers didn't notice us, which seemed dangerous, but could be used to our advantage. For example, I could pass someone who typically wouldn't allow another vehicle to pass them. My driving became a little questionable, as I couldn't help reverting to some old tricks I had learned as a motorcycle courier in London many years earlier.

Speed traps and traffic light cameras typically took pictures of the front of the vehicle, and, of course, our licence plates were on the rear. Footpaths and opposing one-way streets were now legitimate routes, and parking was the easiest of all. We could simply pull the bike up wherever we liked, lock it to something solid, and walk away. As reckless as this seemed, it was the norm here, and the more aggressive the riding, the more acceptable it appeared to be.

Even the police looked the other way from what some bikers got up to.

From where we had camped, reaching the centre of Paris took just over an hour. We parked my bike on a side street next to the Eiffel Tower and wandered beneath the intricate lattice of steel that forms one of the world's most recognisable landmarks. Hordes of excited sightseers formed orderly lines, a rare sight in France, hoping to scale the heights of this iconic structure. We strolled over the Seine River to the fountains of the Jardins du Trocadéro, dipping our feet in the cool waters while enjoying the last of the day's sun. We knew the sunshine wouldn't last.

Parisian weather never could make up its mind.

With only a short time to savour the sights and sounds of Paris, we were drawn back to the river. As we walked along its banks, the setting sun turned the sky into a palette of pinks and reds, all reflected in the calm waters of the Seine. Later, we meandered along the Champs-Elysées towards the Arc de Triomphe where we stood, mesmerised, as traffic hurtled around the impressive monument, seemingly gliding over slick cobblestones, not a traffic light in sight.

As darkness fell, we returned to the Eiffel Tower in time to catch the hourly light display that brought the steelwork to life in a sparkle of a thousand glittering, bright flashes. As we watched with countless others, a full moon peeked out from behind a cloud and completed the show. We left the city centre in great spirits, already planning our next visit, undaunted by the hour-long drive back to camp.

Somehow, that single hour turned into *four*.

After enduring a forced tour of every suburb in Paris, a torrential thunderstorm of lightning, rain, and emotions, and almost running out of fuel, we finally arrived back at camp, exhausted and thoroughly soaked, vowing never to return to Paris for as long as we lived.

The next day, when we paid another visit to the computer store and realised we would be stuck in the area for a few more days, we reluctantly decided to give the city another chance and moved to a campsite in Paris. With an early start, we made the most of a full day in the city's historical centre. Again, we wandered along the Champs-Elysées, one of the world's most famous commercial streets. We hadn't noticed on our initial visit how the entire avenue appeared overrun with temples to consumerism. Sarah and I felt unsettled to see a street with so much historical significance reduced to an open-air shopping mall. I wondered if the local McDonald's had considered adding an extra arch to the Arc de Triomphe and painting it gold.

Those around us who were not in a shopping frenzy shuffled along with the glazed eyes of smartphone zombies, texting and tweeting their every move, taking arm's-length selfies of their heads obscuring the city's most beautiful landmarks. While people crowded the footpaths, there was an eerie hush around us, not at all congruent with the vibrant city we remembered.

This version of Paris was not what we had come to see. Following our ears and noses, we soon found where the locals gathered. Loud, passionate conversations filled the maze of side streets, and the smell of food had our taste buds tingling. I'm not sure what we had expected to find, but this seemed to fit the bill. In our oil-stained motorcycle gear, we were undoubtedly out of place among the crowds of well-dressed Parisiennes, but we joined in the cheer, absorbing the atmosphere. All along the banks of the Seine, locals gathered to celebrate nothing more than the end of their working day, enjoying overflowing hampers of fresh food and fine wine.

Now, *that* was more like it.

***

Content that we'd given Paris another chance, we were also fully aware that after four days of wandering through the city, we'd still barely scratched the surface. When we received the news that our laptop was in good working order, we considered how we were falling further and further behind in our rough schedule. We agreed that we would certainly return for more of Paris, but for now, we had to leave the region and make our way towards Spain. Having never experienced the Atlantic coast of France, we chose a route that took us west before turning south towards the Pyrenees.

Unbeknown to us, a series of powerful thunderstorms had made similar plans.

After several days of blinding rains and exhausting crosswinds, our resolve crumbled, and we booked into a cheap hotel in the hope of drying out our gear and getting a full night's rest. It was our first night in a real bed in over two months, and as our dripping gear formed multiple puddles around our room, we watched a French news channel reporting on the path of destruction left behind by the unseasonable weather.

The following morning, the weather looked promising, so with fresh spirits and dry gear, we loaded up and took to the road. But within thirty minutes, we were completely drenched again.

C'est la vie.

This pattern repeated for several days, but gradually, the further south we went, the warmer it became, and by the time we reached the Bordeaux region, blue skies began to dominate the grey. We celebrated our first dry day with a fine bottle of the local grape juice. Before long, we entered the foothills of the Pyrenees, leaving behind the flat, straight roads of the coast.

Careful route selection took us well off the beaten track, and a few wrong turns brought us into deserted mountain villages. However, the rewards were worth it. Smooth, snaking ribbons of asphalt took the square edges off our new tyres. The warmer air penetrated our protective suits and drove out every drop of moisture. The faster we rode, the drier we became. For the first time in many days, we arrived at our destination without feeling like soggy prunes.

We entered Spain along a centuries-old pilgrimage route. Our journey took on a timeless quality in the shadows of the Pyrenees, where past and present existed in seamless harmony. We felt some kind of magic envelop us and knew we couldn't leave without basking in it. We decided to settle for several days outside the beautiful Basque city of Donostia-San Sebastián. The comfort of the sun on our skin, as well as the warmth of the local people, convinced us to stay.

The climate there in July was mostly warm with soft winds and the occasional light fog that would disperse with the breeze as the day went on. The evenings often brought a gentle rain that washed the sand from the picturesque bayfront promenade back onto the beach. I thought about the messy storms we'd ridden through and how the rain got into the cracks of our gear, leaving us chilled to the bone.

I hoped that if Sarah and I ever had a conflict that felt like we were caught in a storm, it wouldn't seep through the cracks and make us cold and distant.

CHAPTER EIGHT

The End of the Known World

Personal Diary Entry: 17th August 2013

Day 96 (of 549)

13,264 km of 62,840 km (8,242 mi of 39,047 mi)

The distances we have travelled and those that lie ahead can sometimes overwhelm us. Yet, lying in a tent at night, gazing through the fine mesh at the stars overhead, we are humbly reminded of how insignificant our journey is in the grand scheme of the cosmos.

Though short, our stay in Donostia-San Sebastian was productive and relaxing. One of Spain's most popular vacation destinations, the Bay of Biscay, held azure waters that met golden sandy beaches filled with bronzed locals of all shapes and sizes.

We had planned to work on the bikes, but it turned out to be a bank holiday, so we made our way to the ocean for what ended up being one of the hottest days of the summer. After our recent weeks of inclement weather, we were ready for a change, so we joined the bankers and had a much-needed day off. Sarah and I peeled off our motorcycle suits and unveiled our pasty white bodies to overdose on vitamin D. In less than an hour, we added a healthy shade of lobster pink to the colours on display before retreating to a more shaded spot on the beach.

It felt good to be in Spain, where the pace of life was a little more relaxed than that of France or Germany.

Spain's Donostia-San Sebastian, so close to France, is located in the Basque region, where the locals consider themselves neither Spanish nor French. There is a strong desire for autonomy and a palpable affinity with the Irish, which made for an enthusiastic reception and much interest in our bikes and our journey.

The bikes were due for an oil change, and Sarah's rear brake pads were getting thin, so we spent part of a day sourcing supplies and completing the necessary maintenance. We felt rather proud of ourselves for completing these most basic tasks in a foreign country while working around the daily siestas which were observed religiously. We also enjoyed practising our Spanish, a language we felt more comfortable with, even though we were still far from fluent.

Before we knew it, we were packing our gear and moving on. We spent the next week traversing down the north coast of Spain. The Atlantic coast was full of sleepy fishing villages and deserted beaches, and we often ended our day sleeping to the sound of the rhythmic rush of ocean waves. Up in the mountains that hugged the shore, shepherds kept up the ancient rite of guiding thousands of sheep across northern Spain. Farmers worked their land by hand, rarely paying us any attention as we passed.

It felt as though we had travelled back in time.

We couldn't believe that just recently we were surrounded by tech-gadget-clad tourists in the cacophony of Paris. The lush countryside that, at times, looked completely untouched, and other times, peppered with vineyards and rustic homes, transported my mind to a place of complete peace. As weathered as the locals appeared, smiles came easily, and a genuine benevolence lay just below the surface. An overwhelming feeling of contentment emanated from them, a sense that if they couldn't complete a task by sunset, that was okay – perhaps it wasn't so important.

Our route took us further west, following the contorted coastline alongside the Atlantic Ocean. A couple of Dutch bikers we'd met in San Sebastian had suggested we detour through the Picos de Europa

National Park, so we turned inland and upwards for an additional 200 kilometres of driving bliss. The hot tarmac gripped our tyres and allowed us to push the bikes – and our abilities – to their very limits.

Such roads could be exhausting, requiring all our concentration as the stunning views competed for our attention. We pulled over several times to allow our heart rates to return to normal. At one rest stop, we encountered the very same Dutch bikers who recommended that detour. We took the opportunity to thank them for their advice and share a few tales from the road before completing the circuit, which absorbed most of our day and was well worth the effort. As we pulled into camp, drenched in sweat and buzzing with adrenaline, we savoured an ice-cold beer before settling down for a great night's sleep.

***

The next day, pilgrims were everywhere.

Our route took us alongside the Camino de Santiago, an ancient pilgrimage route to Santiago de Compostela, where hikers often walk over 800 kilometres to reach the holy city. The closer we got to the city of Santiago, the more pilgrims we saw as several alternate routes converged for the final leg of this impressive undertaking.

The city was beautiful, narrow streets creating a labyrinth around a huge cathedral dominating the central plaza, offering a gathering point for exhausted hikers, each contemplating the significance of their achievement. Some wept, some chuckled, and some sat silently, staring at the impressive edifice that towered over them. We felt a little like impostors, our efforts overshadowed by this collection of ragged travellers. We wondered what our emotional state would be when we reached our final destination.

The end still seemed so far away.

We spent two days in Santiago, wandering the streets and exploring the markets before another storm prompted our hasty departure. Southern regions promised better weather. Other fellow travellers had suggested we explore the roads of Eastern Portugal, so we pushed inland and crossed another frontier.

As soon as we entered Portugal, we noticed several changes. Although fuel prices climbed, just about everything else became cheaper. There were far fewer vehicles on the roads, and signage gave us more of a general idea than an actual direction. The attitude was, "If you've made it this far, you can probably figure the rest out by yourself."

Overall, the country was rougher, more rural, less manicured than where we had come from. The people were no less hospitable and appeared happy to welcome visitors, even though we were on the less-travelled eastern side. The road numbers on our maps rarely corresponded to what we found on the ground, but we managed to locate the ones we'd been pointed towards. We soon forgot all the frustrations and wrong turns as we wound back the throttles and snaked our way into the mountains.

Oh, the joy of the open road.

The empty roads of the Serra da Estrela mountain range allowed us to thoroughly enjoy every corner without getting stuck behind any oversized camper vans – or road maggots, as we affectionately referred to them. We thought we had ridden the best roads in Europe when we criss-crossed the Alps, but this region was on par. Add to that the warm weather and the lower cost of almost everything, and we had a winning combination. We came in under budget every day in Portugal, helping to offset our overstay in Switzerland. Just when we thought it couldn't get better, we stumbled on three exquisitely beautiful villages and the best campsite of our trip.

We rarely learned from our mistakes and had spent yet another day cramming in too much. We were confronted with the undeniable reality that time and distance possess a peculiar elasticity, defying the laws of physics when one opts for the least direct route with the most hairpin bends. We had to abandon our intended destination, so we settled for following signs that promised a campsite near the village of Beirã. As the roads narrowed and the surface deteriorated, we assumed we'd missed a turn and were just about to backtrack when we stumbled upon the Beirã-Marvão Campground. The Dutch owners instantly made us feel like old friends.

After a long talk with our hosts over a bottle of local red wine, we settled into our tent later that evening. We left the rain cover off and marvelled at the night sky sparkling through the transparent mesh, taking comfort in each other's arms.

"Do you ever get overwhelmed by it all?" Sarah whispered. "Doesn't it feel a little surreal at times?"

"Of course I do," I gently replied. "There are days when I open our maps, and the scale of this journey makes me dizzy. Are you having doubts?"

"Sometimes . . . " Sarah propped herself up on her elbow, seeming pensive. When she spoke again, I heard the uncertainty that had crept into her voice. "I find myself wondering why we are doing this. Like, what's the point?"

When I said nothing, she sighed, sounding more relaxed than she had been a minute ago, and lowered herself back down, nuzzling into my chest. "And then we have moments like this," she continued, her voice soft and warm again. "Moments where everything just seems so . . . flawless. Like we are exactly where we are supposed to be, and I wouldn't change a single thing."

"You are not alone," I assured her, my hand reaching out to cradle her head. "I have those same doubts too. If at any point we agree that we want it to end, we'll make that happen." I began to stroke her hair. "This adventure isn't more important than our own journey – together. I hope you understand that." I paused, feeling the texture of her silky strands between my fingers. "Every day I watch you on your bike, and I wonder how I would cope if I were to lose you."

Sarah pulled away from my chest and I sensed her angling her head as if to look at me inquisitively, although neither of us could see the other clearly in the dark.

I knew what I needed to say, so I kept going. "I realise the stress that creates can sometimes make me a little grouchy, but I want you to know I am here for you. I always will be." I kissed her forehead, wanting all of her uncertainty to dissipate. "I know this trip has sometimes been hard, but it's these magical moments, in between all the challenges, that make it all worthwhile."

"It's easy to lose focus on what we have, isn't it?" she said as she pulled me closer.

For some odd reason, I felt a lump in my throat, so I remained silent.

"It's just you and me, always and forever," she sleepily mumbled her favourite sweet-nothing.

"Always and forever," I whispered back to a now-unconscious Sarah.

After that blissful night under a tremendous, starry sky, we arose thoroughly rested. After packing up our camp, we refuelled on delicious coffee, put on our heavy boots, and set out to explore three nearby towns on our host's recommendation.

The towns of Castelo de Vide, Marvão, and Elvas were simply fascinating. Containing the narrowest, steepest streets we had encountered yet, some were only accessible by motorcycle, and a beautiful castle topped each one. Sadly, the town of Elvas took us so close to the border with Spain that we decided to leave behind the empty Portuguese roads and keep our momentum moving south.

After driving through most of the afternoon and evening, we needed rest. With the heat increasing and our patience waning, we regretted leaving the search for a campsite to the last minute. After several false leads, we pulled up outside the town of El Coronil, and too exhausted to continue, we settled down for our second night of wild camping. Hot and sticky and too tired to care, we lay on top of our sleeping pads and prepared for a fitful night's sleep, but sleep nonetheless.

At exactly midnight, the town erupted into a noisy festival of lights and music, which lasted until six in the morning. Clearly, they had heard Sarah and I were in the area and decided to throw a party just in case we should show up.

In the morning, we groggily departed the town and the party's aftermath and set our sights on Gibraltar, stopping briefly for breakfast in the stunning town of Ronda before climbing our last mountain pass as we neared the coast. A rest stop on one of the summits provided us with a view that nearly had us tearing up.

We caught our first glimpse of Africa.

Through hazy ocean air, we could just make out the distinctive shape of the Rock of Gibraltar and the mysterious Atlas Mountains beyond. Motivated to see more, we picked up the pace, and by lunchtime, we were approaching the border between Spain and this old bastion of the British Empire. Security appeared a little tighter here than on any other European border we had crossed yet, with passport checks on both sides of the frontier.

Guarding the entrance to the Mediterranean Sea and dominating the skyline was the actual "Rock" itself. A giant limestone monolith almost half a kilometre high, it has held a significant strategic value for centuries. During World War II, it was shrouded in secrecy while home to over 30,000 British soldiers, sailors, and airmen. A warren of more than fifty-two kilometres of tunnels, chambers, and galleries penetrated deep within, housing bunkers, gun emplacements, barracks, and even hospitals for the wounded.[4] We tried to ride our bikes to the top only to find it prohibitively expensive, so we turned back and decided to return on foot the next day.

The following morning, with hiking boots and packed lunches, we arrived again at the base. We climbed out of the overcrowded city to the top via the Mediterranean Steps, an abandoned, exposed pathway on the south side of the hulking monolith. A troop of slender Barbary macaques, referred to by the locals as "rock apes," quietly observed our progress. Hiking up from the city's sweltering heat below, we soon enjoyed a cool ocean breeze and a spectacular view across the Straits of Gibraltar to Morocco.

Through the distant clouds on the African side, we could see Mount Abyla arise, which, with the Rock of Gibraltar, formed the Pillars of Hercules, once considered the end of the known world in ancient Greek mythology. Gazing out across the vast expanse of the Atlantic Ocean, it was easy to imagine why. Sadly, the Rock had become an overdeveloped tourist attraction, and at every turn, there was a charge or a fee and an orderly line of people willing to pay.

Returning to the city by late afternoon, we rewarded ourselves with lukewarm beer and watched a parade of lobster-red English tourists march up and down Main Street, shopping for familiar goods they could easily find at home for half the price. There was excitement in the air as tensions over fishing rights in the surrounding waters and the bigger question of sovereignty brought the Spanish and British governments into another heated exchange.

As we prepared to leave, we couldn't help but notice the forlorn faces of off-duty soldiers watching the world go by from the tiny windows of their residential barracks. They looked more like they were serving a prison sentence than fulfilling a vocation. One fastidiously patriotic older lady explained to us that "as long as the apes remain, the Rock will always be British." We thought it a rather odd way to refer to your fellow countrymen.

With Africa so close, we could only wonder about what adventures we would encounter ahead. The potential hardships awaiting us would likely magnify all the minor difficulties of our trip up to this point.

CHAPTER NINE

How Life Should Be

Personal Diary Entry: 5th September 2013
Day 115 (of 549)
17,358 km of 62,840 km (10,786 mi of 39,047 mi)
The three-month mark went by unnoticed. As we learn to embrace
the freedom that comes with a life on the road, we can't help but
feel gratitude for how we are progressing. We've had our ups and
downs, good days and bad but, as yet, no major catastrophes.

For a moment, it all felt familiar. I looked at the map and saw a
mountain range called Sierra Nevada, like in California – except those
Sierra Nevadas were over 9,000 kilometres away.

From Gibraltar, we turned east towards Granada, leaving behind the
muggier coastal climate for the cooler, dry air of Spain's Sierra Nevada
mountains. What looked like a beautiful, winding coastal road on the
map, from La Línea to Málaga, turned out to be an overdeveloped
strip of endless resorts with countless billboards promising to fulfil
our every desire.

After living in California for so long, we had considered ourselves
immune to this type of marketing, so after a lunch of thirst-quenching
Coca-Cola and a delicious, juicy Big Mac, we turned inland onto one
of the most spectacular roads we had ridden to date.

The A-4050 climbed steeply out of Almuñécar, piercing the southern side of the Sierra Nevada mountains, quickly gaining altitude with each hairpin bend. A well-maintained surface and little other traffic had us pushing the bikes to the very limits of their intended use, utilising every inch of tarmac, and leaning into turns – sparks flew from the underside of Sarah's bike as she tested her nerve and ability.

I rode behind Sarah, balancing equal measures of fear and admiration.

On entering Granada, we began a fruitless search for the campsite on our map, leaving us hot, thirsty, and tired. We finally stumbled across a different site, on the city's outskirts, where the camp host welcomed us with the sad news that they were full for the night – a brief Spanish holiday was in full swing.

Disheartened, we turned to leave. On top of the reception desk, I noticed a local newspaper opened to a page reporting on the territorial dispute over Gibraltar. I expressed my empathy with their situation and, as an Irish person, my understanding of their frustrations. All of a sudden, a reshuffle took place, and a modest space became available. After pitching our tent, we satisfied our thirst with an ice-cold beer while raising our glasses to British foreign relations.

***

The central purpose of our visit to Granada was to explore the Old City and the Alhambra. Eight centuries of occupation by the Moors had left behind an indelible imprint of the Islamic culture. The contorted, narrow streets were home to many a tea house with low tables, lush rugs, and plentiful hookah pipes. In the Sacromonte district, gypsy caves were carved directly out of the rocky hillsides.

Quiet and reserved during daylight hours, the streets came alive after dark with the sultry tones of classical guitars and the provocative moves of the fierce Flamenco dancers. The nearby gritty Albaicín district provided an impressive view of the Alhambra fortress while its winding alleyways brought to mind its mediaeval past.

From within the Alhambra complex, the ancient castles of the Nasrid dynasty overlooked the city of Granada. The fortress walls, so plain on the outside, belied an exquisite interior of remarkable beauty and intricate detail. Every carved tile told a story, and combined, they formed a poetic chronicle, exalting the achievements of a long-dead empire.

It was paradise.

Or, at least, that's what they were going for when they built the place. The palace blended running water, sunlight, open air, and subtle symmetry to create a peaceful oasis while concealing traces of a past tainted with intrigue and bloodshed. We spent a long, hot day exploring the site before returning to the Albaicín district as night fell to enjoy some tasty tapas.

All too soon, we were back in the saddle, moving north through the Sierra de Segura mountains and lakes before pulling over at Alcaraz for the evening's camp and the best paella we'd ever tasted.

Torn between the mountains and the ocean, we moved towards Andorra, at times following the coast before seeking the cooler inland air of the higher elevations. Campsites, while more common the closer we got to the Mediterranean Sea, started to resemble large-scale holiday resorts, replete with scheduled events for the whole family to enjoy.

We pulled up at one campground and asked for their cheapest site, only to be told it would cost us fifty-five pounds for a tiny piece of dirt and likely a tepid shower. I enquired why it was so expensive and the receptionist assured me, "Because it is the 'high' season." When I asked her if she was high, my sarcasm was, fortunately, lost in translation.

Turning our bikes around, we discovered a perfect wild camp nearby and wandered back later to use their facilities for free. We would've considered staying there for half the price but didn't enjoy getting ripped off. The area seemed to smell of money – and also desperation. This stretch of Spanish coastline felt, at times, overdeveloped and, at others, totally abandoned. Lavish hotels and vulgar luxury villas appeared at regular intervals, contrasting with the barren, dusty landscapes separating each resort. Sex workers were a common sight on the outskirts of each city, where they sat quietly under red umbrellas, wearing silk and lace over tired bodies.

As always, when a place heard that Sarah and I were passing through, it knew how to welcome us. We thought we had slipped beneath the radar as we pitched our tent on the outskirts of the small town of Áger. But, at exactly midnight, the town, which couldn't have been home to more than 1,000 people, erupted in a festival of dance and music, which continued until six in the morning. We lay awake in our tent for most of the night, unsure whether to ignore the commotion we heard through the paper-thin fabric or give up and join the festivities.

***

Leaving Áger, we took a spectacular road northeast into the tiny principality of Andorra, a unique nation famous for its skiing and shopping. The twisting mountain roads and extensive ski resorts, made us feel like we were back in Switzerland, which we still fondly recalled as "maybe the place." We couldn't resist staying for a couple of days to explore the area.

Andorra had only three major roads, so it took less than a day to ride them all. A vague network of hiking trails allowed access to some of the most spectacular scenery in the Pyrenees mountains. Before long, we were puffing and panting our way up into deserted valleys, towards exposed summits of loose rock and chilling winds. As we approached the top of one ridgeline, I noticed a large bird soaring overhead, circling directly above us, taking a particular interest in our every step.

"Is it just my imagination, or does that vulture seem to be checking us out?" I asked Sarah.

She laughed.

"When we were in France all those years ago with my family," I continued, "my brothers and I went bird watching. I distinctly remember the guide describing a large vulture that looked a bit like that, which lived in the Pyrenees. It would knock unsuspecting animals, and sometimes even shepherds, off the ridgelines so it could feast on their tenderised corpses below."

"It wasn't called a Lammergeier, was it?" Sarah said, sounding startled.

"Yeah, that was it! How'd you know that?" I replied, impressed by her knowledge of the local birdlife.

"There was a notice about it at the beginning of the trail. I couldn't decipher the whole thing, but it appeared to be some kind of warning." She had stopped in her tracks and was nervously eyeing the bird.

"Do you think – ? No, it can't be, although it does look like . . . " My words trailed off as we stepped back from the edge of the ridge.

As I put a protective arm around her, she put her mouth near my ear and whispered: "Well, I think you'll be safe."

"Why's that?" I asked.

"Have you caught a whiff of yourself recently? If it wants fresh-smelling meat, you're not it!" she chuckled as I loosened my embrace.

"Yeah, fair point, I guess," I muttered under my breath, making a mental note to seek out a campground with a shower soon.

***

From Andorra, we descended into France, determined to give the Mediterranean Coast another chance, only to find the region hideously overpriced and overdeveloped. Turning the bikes inland again, we discovered many hidden gems, small towns surrounding impressive strongholds, delightfully empty roads, and impressive landscapes.

That terrain was more our cup of tea – or rather, glass of merlot.

The Verdon Gorge provided us with plenty of thrills, the narrow road clinging to the vertical walls of the yawning canyon, castles standing out on distant hilltops, and vertigo-inducing drops just beyond the abrupt edge.

The most direct route along the coast, from France into Italy, took us through the tiny principality of Monaco. A tax haven, it was, coincidentally, home to some of the wealthiest people on the planet. I couldn't help but wonder what kind of world would be possible if the rich actually paid their fair share of taxes.

Nestled between the Southern Alps and the Mediterranean Sea, Monaco occupies a spectacular coastal area, which it does its best to spoil with high-rise apartment blocks and gaudy casinos. The constant gridlock of this compact state smothered the true potential of the extravagant, luxury sports cars that prowled the streets. Superyachts crowded the bay, some too big to dock in the harbour, while helicopters shuttled their owners to and from the mainland.

After half a day marvelling over the excesses of this unsettling country, disturbed and appalled by the grotesque displays of wealth, we rode east towards Italy, gracefully zipping between the growling Bugattis and Bentleys that choked the narrow streets.

Italy welcomed us like we were family.

Throughout history, Italian communities have gathered in their piazzas to celebrate the end of wars, worship saints, or protest for their rights. They also gathered just because, without needing any reason other than they're Italians, and it's better to eat and drink with good company. The same sense of community still lingers centuries later.[5]

Our first night at a campground off an obscure road north of Ventimiglia turned out to be one of our most enjoyable evenings yet, as a wonderful local couple introduced us to a quintessential Italian dining experience. Once a week, the townsfolk would fire up their large communal brick oven and turn out pizzas, two at a time, topped with whatever was in season to feed the crowds of hungry diners who bided their time with good conversation and smooth wine. We had six nationalities at our table that evening and little difficulty communicating.

Our new friends, over homemade cappuccinos the following morning, recommended several routes for the next leg of our journey, and, seduced by the prospect of more winding mountain roads, we turned north on the 6204 towards the province of Cuneo.

The road meandered between France and Italy while climbing towards the Col de Tende mountain pass. We stopped for lunch in the tiny town of Tende, wandering through its quiet streets, intrigued by the network of alleys and passageways that seemed to follow no logical pattern.

We had set our sights on the town of Alba as our goal for the day, and as the road gradually returned to the foothills, we entered one of Italy's wine-producing regions. Vineyards stretched into the horizon and the afternoon clouds had us stopping at every turn to capture the magical landscapes with our camera.

Soon, these same friendly clouds, which had looked so grand from afar, were overhead, taking on a more threatening appearance. As the sky darkened, the first few plump drops of rain burst upon us, and we sought shelter beneath an abandoned gazebo next to the road. No sooner had we pulled our bikes out of the rain than an immensely powerful lightning storm lit up the horizon.

We decided to wait it out.

Dry and loaded with all the supplies we needed, we cooked a tasty meal on our camp stove while lightning bolts scorched the surrounding air. After a couple of hours, when it appeared the storm had settled overhead, we relented and booked ourselves into a hotel for our second night in a real bed since entering Europe.

***

Well-rested, we once more made our way south and picked up the coastal road towards Pisa. The previous night's storms had washed out several sections of the smaller local roads, and a lack of clear signage had us backtracking and following our instincts on several occasions. Finally, we pulled into Pisa and enjoyed a late evening walk around the surprisingly small historical centre. As we gazed upon the iconic leaning tower, a nearby church choir sang beautifully, their voices carried gently upon the light evening breeze. Far to the north, another thunderstorm flashed and flickered behind thick, ominous clouds, silhouetting the entire plaza.

Turning inland the following morning, we arrived in Florence, amazed to discover that Italy could get more beautiful. We wondered if *this* was the place where we could end our journey. The people, the architecture, the atmosphere, and the food all combined to give this country a unique flavour of how life should be. There was a noticeable passion in the simplest of events – even a conversation could feel like a performance. We attracted many excited enquiries about our motorcycles and our journey, as well as invitations to dinner and advice on the best roads to take. We thought it couldn't get any better – but then we arrived in Rome.

"The Eternal City" was unlike anything we had ever experienced.

Every turn revealed a glimpse of an impressive history spanning over 2,500 years. Monuments were everywhere, some in ruins, some restored, each with a fascinating story. We often eavesdropped on the guided tours and picked up snippets of captivating detail.

As we approached St. Peter's Basilica in the Vatican City, we noticed thousands of people slowly leaving the area. As it turned out, we had just missed the celebration of Sunday Mass – old habits from my youth die hard, I guess. The scale of the event and its location were hard to fathom. All the places of worship we had seen up until now were dwarfed by this, the very heart of the Catholic church.

It was also quite disturbing.

Wandering the floors of the Vatican Museum, with its overwhelming display of artwork from throughout the ages and gift shops at every turn, it left us questioning how the church reconciles this hoarding of priceless treasures with the original teachings of its founder. We could only wonder at what a difference such an abundance of wealth would make if used to ease the lives of so many in much greater need.

Within the Basilica, the murmur of hushed voices accompanied the sound of shuffling footsteps on a tiled floor polished smooth over centuries by reverential pilgrims. Occasionally, the creak of an ancient, heavy hinge or the toll of a solemn bell would echo throughout the vast chamber.

As I gazed upon the endless oil paintings from a bygone age, many illustrating dubiously enlightened visions of the wonders of heaven or the horrors of hell, they reminded me of a habit I had formed many years before.

Growing up in Northern Ireland during a period of bitter conflict and division left me deeply suspicious of organised religion and acutely aware of my mortality. Indoctrination from a young age resulted in more questions than answers, and I often doubted the existence of a higher being or an afterlife. My faith was uncertain, and I slowly drifted away from believing everything I was taught, but my scepticism had flaws too.

What if I was wrong?

As our hobbies took my friends and me on a path towards riskier behaviours and more dangerous pursuits where survival wasn't always guaranteed, we coined the rallying cry, "First pub on the right!" just before reaching that point of no return.

It morbidly implied if one or more of us didn't make it out alive, and we found ourselves in a heaven – or, more probably for us, a hell – we could find solace in knowing we would meet again at that location: the first pub on the right.

Meandering through the church's cavernous corridors, I carefully studied each of the artworks that adorned the walls. There were paintings and sculptures of lovers, weddings, and babies. I inspected portraits of Adam and Eve, Jesus, his mother, and angels of various ages and sizes. I saw beautifully rendered soldiers, saints, and martyrs with their intestines torn from their bodies. Walking from room to room, I gazed at illustrations of churches, tombs, and prisons.

I was a little disheartened by the complete absence of pubs in any of them.

CHAPTER TEN

Without a Destination

Personal Diary Entry: 18th September 2013

Day 128 (of 549)

18,491 km of 62,840 km (11,490 mi of 39,047 mi)

Abandon all maps, ye who enter Venice . . .

Our whirlwind tour of Rome took us to countless wonderful sites and, no doubt, we ignorantly strolled by many others. What little preparation we had made before our visit would never be enough – Rome would take a lifetime to explore and many more to fully appreciate.

On the morning of our fourth day, we each tossed a coin over our shoulders into the Trevi Fountain, hoping to ensure our safe return, as the custom dictates. From there, we turned north towards the ancient city of Assisi, in the Umbria region, east of Tuscany. An endless blanket of vineyards and olive groves covered the rolling landscape, with Roman forts and mediaeval settlements capping the peaks on either side of our route.

Upon arriving at a campground near Assisi, a very kind Italian couple immediately adopted us and insisted on feeding us in exchange for some tales from our journey and a few songs on my guitar. After such a wonderful welcome, we decided to stretch out our stay and rent a couple of bicycles to explore the pristine city at a more leisurely pace.

87

The steep, cobbled streets and immaculate buildings displayed little evidence of the earthquake that devastated the region in 1997.

If anything, it all appeared a little too perfect.

With the reconstruction so recent, it had little chance to weather and age like other towns of that period. Finding the ancient city a bit too modern, we spent more time getting to know our new friends at the campground.

On the third day, we bade them a sad farewell and made our way indirectly to San Marino, the world's smallest republic. While taking a maze of rural back roads, we crossed the border without even realising it. We were soon winding our way up the side of Monte Titano and into the capital, where the Three Towers of San Marino overlook this Lilliputian country. With a population of approximately 33,000, the Sammarinese enjoy one of the highest life expectancies in the world.[6] I suspected it could have something to do with the exercise they got from their steep streets and endless steps.

After a short stay in San Marino, we coasted downhill towards the Adriatic Sea, determined to complete our Italian experience with a visit to Venice. A Polish-Ukrainian couple we'd met in Rome had recommended a campsite to the east of the city, so our route curled around the north side of the marshy Venetian lagoon before turning back along the peninsula to Punta Sabbioni.

Our experience in Venice contrasted sharply with that of Rome. We had squeezed too much into our limited time there, and I had grown frustrated with our inability to see it all. We each had an agenda about where we wanted to go and what sights to enjoy. It had led to some fissures between us as we stubbornly dug in our heels over what the plan for each day should include. Neither of us wanted to compromise, so our connection suffered. I disregarded Sarah's desires while selfishly pursuing my own.

I knew I wasn't always the easiest person to be around.

After a decade together, our shared history fostered complacency, and I occasionally took our relationship for granted while not being sensitive towards Sarah's wants and evolving needs. Although I acknowledged this tendency, summoning the energy for transformative change wasn't easy. Confrontation seemed daunting, prompting me to seek solace in distractions rather than addressing our problems head-on. I became entangled in an unhealthy paradox; as I side-stepped the inevitable, I inadvertently heightened the underlying tension.

In the quiet moments inside my helmet, as we covered the vast distances between our destinations, I contemplated our relationship and the challenges we faced – some real, some imagined. I resolved to try harder.

So, in Venice, I suggested an idea.

Our camp, near Venice, was an easy walk from a ferry terminal which ran regular shuttles between some of the 118 islands that make up the city. Fellow travellers had forewarned us to expect to get lost in the mischievous maze of corridors and canals that carve up the small city. To avoid this confusion, I brought up the option of simply abandoning our map and wandering aimlessly.

Without a destination, it is impossible to become lost.

As we dodged flustered sightseers standing at crossroads and spinning their maps this way and that, we stumbled upon some of the less-visited parts of the city. We became entangled in a delightful game of hide-and-seek within the elusive alleyways. Locals occasionally gave us odd looks, wondering what we were doing in their part of town, far from most tourists. One resident we spoke with, who'd lived there for several years, joked about how she would discover a new street each time she went from A to B.

Like carefree children, we chased each other, took unexpected rights and lefts together, and stopped suddenly before grabbing hands and slowing to a stroll to casually discover what was around this bend, behind that chapel, across the canal. Exploring Venice that way, on foot and with no daily schedule to keep, helped us feel grounded and relaxed, captivated by everything we saw.

Once a major maritime power, Venice's former extravagant wealth was still evident in the grand, yet crumbling, remains. We weren't wealthy but we were rich with experience as we followed the sunlit stones of a path leading to a canal, intending to have a simple, luxurious meal right beside the water. Once we found the right spot, with the best reflection of the changing colours of the sunset sky, we spread out a picnic and sat side by side, watching intricately detailed gondolas drift by with romantic couples aboard.

It would be difficult to imagine such a floating city if it didn't already exist.

Olives, cheese, bread, and tomatoes washed down with a local red wine provided a fitting last supper for our time in Italy. As we gathered our belongings, a plump rat scampered over our feet before splashing into the narrow canal and swimming off towards a sagging house upstream. Sarah shrieked and I laughed as we walked back to our ferry, ready to move on from Venice.

CHAPTER ELEVEN

Off the Beaten Track

Personal Diary Entry: 26th September 2013
Day 136 (of 549)
19,308 km of 62,840 km (12,068 mi of 39,047 mi)
At the end of the day . . . it is nighttime.

No matter how much we try to control or influence our lives, we
are ultimately subject to the natural laws of the universe. Time
waits for no one.

New terrain and a challenging language welcomed us as we crossed the
frontier into Slovenia, a country we knew little about. The early signs
of autumn were beginning to show on the vast expanses of forest that
covered over half of this republic, providing a subtle reminder of how
long we had been on the road. As our route climbed into the Julian
Alps, a significant drop in the temperature accompanied a light drizzle,
and we added extra layers of clothing to counteract the chill.

Soon, we were lost, and all our extra layers were soaked.

Frustrated, we pulled over to compare our map to the confusing
road signs. Another long-distance biker from Serbia joined us to
take photos and share stories. His positive attitude and beaming
smile quickly made us forget about the inclement weather and our
damp spirits. Suitably encouraged, we pushed onwards and upwards

into the Triglav National Park. The ride dried out most of our clothing, and we comfortably began to set up camp near the village of Bohinjska Bistrica.

A torrential downpour began within minutes of pitching our tent and cooking dinner. After a long night of thunder, lightning, relentless rain, and barely any sleep, we packed our dripping gear and retreated to a warm, dry hostel in the nearby town of Bled. Once we had hung up our drenched clothing and equipment, we collapsed into bed after hardly speaking to each other all morning. We were exhausted.

"I think we should rest today and then, maybe, spend a day apart," I suggested, worried about how she might respond.

We had been on the road for over four months together, and the signs of stress were showing. The terrible weather wasn't helping.

"Maybe that's a good idea," Sarah replied, her voice cold and distant.

I wondered whether my honesty had inadvertently crossed a delicate boundary and caused unintended hurt. I did love her and enjoyed her company. I was worried, though, that too much time together was causing the friction that led us to squabble occasionally. We needed to be there for each other, but not for every second of every day. Perhaps with a day apart, we might miss each other – something we didn't seem to do anymore. Tensions had arisen between us at various times throughout the journey. With more experience in motorcycling and travelling in general, I felt the burden of decision-making and route planning predominantly on my shoulders. Balancing responsibilities and maintaining my composure wasn't always my forte. It could get tiresome, and I wanted to be more adept at controlling my temper.

I was finding it hard to pour from an empty cup.

"So that's decided, then," I said. "Let's give each other a little space today as we catch up on some much-needed rest, and tomorrow, we'll spend an entire day apart. It'll be our first since we started."

I reached out and took her hand as we rolled towards each other on the bed. I looked into her eyes to get a sense of how she was feeling. What is the appropriate way to respond when your partner tells you they don't want to spend time with you? I didn't perceive any resentment or animosity, just weary resignation.

"It's probably for the best," she finally replied as her eyes closed.

Two nights in Bled gave us the opportunity to rest, dry out our gear, and plan the next leg of the journey. We enjoyed some local hikes and sights and spent one day apart as we had agreed. Feeling refreshed and renewed, we acknowledged it had been a wise decision and decided to make it a regular practice.

After getting a glimpse of what Bled had to offer, we moved southwest towards the capital, Ljubljana, intending to briefly explore the "City of Dragons." As we packed our bikes, we noted the date: Friday the 13th. We promised each other we'd take it easy on the road that day. Though neither of us were particularly superstitious, we wondered if we would encounter any surprises or upsets.

As we pulled into the city centre, yet another party appeared to be in full swing. Slovenia was hosting the European Basketball Championships, and the capital throbbed with fans and parties. We strolled through the centre given over to many street stalls selling merchandise and food, sampling some delicious potato and onion dumplings. We hadn't planned to spend too much time there, so by early afternoon, we returned to our bikes and drove south towards the country's border with Croatia.

A pleasant ride through the heart of this beautiful country took us to a border crossing outside Banja Luka where we assumed we could cross without hindrance. With my Irish passport, the border guard waved me through without question. However, he took a special interest in Sarah's US passport, and on further inspection, he noted her only stamp, from Dublin, granted her a 90-day visa. With well over 120 days on the road, he refused to let us go any further, insisting he would have to classify her as an "overstayer" and put a fat red stamp in her passport to prove it. As this action would undoubtedly create issues later, we pleaded our case, and his only recourse was to allow us to turn back and consult our embassies in Ljubljana.

Being late on Friday night, we resigned ourselves to a longer stay in Slovenia than we'd planned, to wait out the weekend before visiting our embassies on Monday. With our initial experience of the city so positive, we returned to the capital to look for a comfortable camp. We were determined to research where we stood on the legal aspect of immigration and visa extensions. It didn't take long to stumble across the Schengen Agreement and its implications for the various EU countries that had signed it.

It was rather complicated.

Each country within Europe could be either EU or Schengen, neither, or both. The Schengen Agreement, designed to ease border crossings for EU citizens, opened trade and commerce opportunities while dissolving the role of official border posts. It was now virtually impossible to get a stamp when crossing a land border within Europe, although each country appeared to have its own interpretation and guidelines for enforcement.

On paper, it looked as though Sarah had overstayed, but she hadn't entered Schengen territory until we rode into France. Both Ireland and the UK had refused to join. We hoped, come Monday, the embassies would help Sarah out.

The weekend passed quickly. We were able to catch a few basketball games with some of the locals, and as Monday dawned, we sat patiently waiting for our first embassy to open. We assumed between the British, US, and Irish embassies, we'd receive some help or, at least, some useful advice. We had carefully checked our dates, and it was just about 90 days since Sarah had entered France, but with nothing to prove it, we hoped for a sympathetic diplomat to throw us a lifeline.

The expression "as useful as a chocolate teapot" came to mind when recalling the helpfulness of our distinguished foreign service employees. The best advice we received was "try to sneak across at a busier border post and hope for the best." After exhausting our options, we realised it was time to come up with our own plan – so we resolved to sneak across at a busier border post and hope for the best.

In the meantime, we had become rather attached to our new friends at the Laguna Campsite on the outskirts of Ljubljana. On our intended departure day, the heavens opened yet again, so we spent the entire day sheltering from the rain, drinking potent "Indian Water" at the bar with an Iranian, an Israeli, a Serb, and a Croat.

There has to be a punchline in there somewhere.

While temporarily delayed in Ljubljana, we decided to take care of some basic bike maintenance. The chain and sprockets on my bike were long overdue for a change. I had stretched their lifespan well beyond their recommended use and had been reminded of this every day by an uncomfortable grinding sensation each time I rolled out of camp.

A visit to the local BMW dealership in Ljubljana proved fruitless, so we tracked down another bike workshop, Avtoval, ten kilometres away in Grosuplje. We made the short journey out of the city and were soon talking to Dejan Valentinčič, who insisted on selling us the parts at a significant discount while fitting them for free. For the first time in many years, we left a BMW dealership feeling like they hadn't taken advantage of us. The mechanics even let us get involved in their workshop in case we had to do those tasks by ourselves somewhere down the road. The staff at Avtoval went above and beyond, lending support to people actually using their bikes for the purpose for which they were designed.

With deep regrets that we couldn't stay longer, we bid farewell to our newfound friends in Ljubljana who plied us with more "Indian Water" on the morning of our departure. After hugs all around, especially an extra-long one between us and our new Israeli friend, Shlomi, we climbed onto our bikes. We adjusted our route to take us east towards Croatia, through Slovenia's second-busiest border crossing.

By early afternoon, we had followed the Sava River to the border. After a cursory glance at our passports on the Slovenian side, we passed through a deserted Croatian border post and cruised on towards Zagreb. We found a camp on the outskirts of the city and spent the rest of the day exploring the capital. We met a friendly German couple who suggested we alter our intended route south to take in the Plitviče National Park, so the following morning, we wound our way along Highway 1, taking in the impressive landscapes of this new country.

Quiet villages lined the road as locals worked the land, harvesting crops by hand. It provided another reminder of the impending end of the summer. Pockmarked walls of shell-shocked buildings subtly marred the tranquil settings, evidence of the conflict that gripped the region not long before.

Plitviče National Park proved to be a worthy detour. Unfortunately, our visit coincided with the weekend, and an endless stream of tour buses converged on the park as we pulled up. Nevertheless, we set up camp for a couple of days to give us time to appreciate the area. We managed to avoid the tour groups that swarmed the most popular attractions and barely kept up with their overwhelmed guides. Instead, we mostly stayed off the beaten track to explore the lesser-visited areas of the park.

The park was beautiful, with a cascade of pristine lakes linked by crystal clear waterfalls, criss-crossed by wooden walkways and narrow paths. We spent an entire day marvelling at the wonderful sights and dodging the crowds before returning to camp and plotting the next leg of our journey.

For as long as I could remember, the coastal highways of Croatia had held a special place among the worthy roads in biker lore. We were curious to see if they lived up to their reputation. We knew we were in for a treat as we came over the last ridge, and the Mediterranean Sea opened up before us with countless islands stretching off over the horizon. We joined Highway 65 at Senj and turned south, hugging the shoreline, the subtle tang of ocean air wafting up from the translucent waters. The temperature rose, and the road surface improved, allowing us to thoroughly enjoy the twists and turns, losing ourselves in the delight of riding this incredible stretch of highway.

We were so lost in the moment that we almost didn't notice the gnarly old police officer in his crinkled uniform waving us over after checking our speed on his radar. Ever the diplomat, Sarah removed her helmet after we pulled over and gave him a smile that instantly melted his steely expression.

He soon smiled with her and asked, respectfully, if we could slow it down a bit.

Branching off the main highway, we crossed onto the island of Pag. Riding to its most-northerly point, we found a camp and were soon settling down to a tranquil night under the stars. Enjoying the warmer temperatures, we slept peacefully, excited at the prospect of riding more of the Croatian coast.

Shortly after sunrise, we packed up and hit the road, passing tiny fishing villages clustered around pristine inlets. Blue skies and azure waters made for a spectacular day's riding, and before we knew it, we had reached our destination.

Again, it felt like we'd travelled back in time.

The city of Split, on the Dalmatian coast, has an incredible centre nestled within its sprawling suburbs. Diocletian's Palace forms the heart of the old town, a beautifully preserved Roman ruin still serving as a bustling hub of shops, restaurants, and accommodation. The polished stone pathways have welcomed countless visitors since the first inhabitants in the fourth century AD, and after the deserted ruins of Rome, it was refreshing to see and feel the complex brought to life. It didn't take much imagination to get a sense for how it must have felt during the reign of the Roman Empire. A few enterprising locals in legionnaire costumes roamed the streets and posed for photos with tourists willing to pay for the privilege, adding to the overall atmosphere.

From our relaxed camp outside the city, we could bus into town, relieved we didn't have to deal with the choked city streets. We spent evenings relaxing on the beach, watching the sun set over the ocean; morning swims in the cool, clear waters made for a great start to the day.

From there, we planned to try our luck at the border of Bosnia and Herzegovina before returning to Dubrovnik soon. Our episode at the Slovenian border had served as a useful reminder of how officialdom could throw a spanner into our plans at any time, with the likelihood of more difficulties as we progressed towards less cooperative countries. We had received plenty of well-intentioned advice, but with the myriad of nationalities and bureaucracies involved, only time would tell how our journey would evolve.

As advised back in Slovenia, we could only hope for the best.

CHAPTER TWELVE

The Vanishing Road

Personal Diary Entry: 4th October 2013

Day 144 (of 549)

20,732 km of 62,840 km (12,958 mi of 39,047 mi)

Despite the seemingly insurmountable distance between our starting point and our destination, I find solace in the realisation that every step we take, no matter how small, inches us closer to our goal. It is a reminder that even the greatest achievements are simply the sum of countless small, incremental victories.

Revitalised after an early morning swim at our campsite south of Split, we loaded our bikes and ate a quick breakfast before planning a route into Bosnia and Herzegovina. The obvious, most direct path was rarely our first choice, so we scanned our maps for the thinnest, most contorted road to take us over the Biokovo Mountains towards Mostar.

As we wound our way out of the deep valleys along narrow switchback roads with hairpin corners, it became apparent that our maps lacked sufficient detail to chart every road. Then, as the surface gradually deteriorated, we soon found ourselves on a loose gravel bed.

The road virtually vanished beneath us.

We should have guessed that the lack of traffic indicated something was wrong. In the distance, we could see an occasional vehicle using an alternative route, so we pushed through and were soon back on asphalt with only a rough idea of where we were.

Before we knew it, the vanishing road had somehow led us to a border post, and as we tentatively approached, we wondered how this one would work out. Sarah told me to cross my fingers in hopes that this crossing would go as smoothly as our entrance into Croatia.

One of the advantages of travelling the way we did was we always had plenty of time and rarely a fixed destination for the day. Our relaxed nature tended to diffuse some of the bluster and authority carried by those in uniform, who often emphasised their importance by making us wait. Unperturbed, we would typically fold our arms and patiently sit until they found an excuse to let us pass.

On this occasion, the Croatian side seemed happy enough with our passports; however, they requested our bike documents. We attempted to communicate with a shrug of our shoulders and hand gestures that we'd have to dig through our luggage. An impatient wave hurried us through, and it was now the turn of the Bosnian officials. A similar response to their requests did not work, and they seemed perfectly content to have us pull over and fumble through our gear for all the necessary paperwork.

Travelling through Europe on motorcycles with US licence plates had often worked in our favour. Parking attendants, speed traps, and traffic police had, so far, left us alone. We hoped our luck wasn't about to run out.

One document we were required to carry was proof of insurance, otherwise known as the "Green Card," which we had been diligent about since day one. However, the insurance companies had conspired to make it as challenging as possible to obtain a single policy that worked everywhere, especially when travelling through Eastern Europe. Our policy was good for most other countries, but not for Croatia or Bosnia and Herzegovina.

It didn't take the border guard long to point this out.

"You have problem," he said.

As we stood there, wondering how to navigate the situation, the guard chortled; his colleagues gathered to look over his shoulder at our documents. They all seemed to be in a rather jovial mood, and we suspected the empty beer cans inside their office may have been a contributing factor.

"Your insurance not good for Bosnia," he added. "Or Croatia. Is big problem."

"We were planning on buying insurance when we found an agent. Do you know where we might find one?" I asked.

This question prompted a long, slurred, and, at times, heated discussion among the guards, which lasted for several minutes. Each officer seemed compelled to add his opinion to the debate. Sarah and I remained silent, awaiting their response. Finally, they all seemed to agree on something, and the guard who had initially spotted the infraction turned to us and said, "No."

We had reached an impasse.

I considered offering to ride to the nearest town to pick up more beer for them, but a distinctive click and hiss from inside the office led me to believe they still had plenty. For a while, they simply ignored us, talking among themselves, and smoking constantly. I couldn't decipher a single word of what they said, but it didn't appear to concern us or our lack of insurance, and we wondered what we should do next.

Sarah whispered, "Should we offer them a bribe?"

"I don't know," I replied.

I had difficulty reading the situation, and I didn't want to add attempted bribery to our lack of insurance problem. We had set out from the beginning with the intention of completing our trip without resorting to offering bribes. We felt it could hinder those who came after us. Once the habit becomes established, the amounts involved would begin to grow and soon, people on a tight budget like ours would no longer be able to afford to travel.

So we waited.

After half an hour, they grew tired of our refusal to become flustered and, returning our documents, told us to go, pointing towards Mostar. We fired up our engines and sped off into Bosnia, grateful.

Yet again, our patience paid off.

Some countries within the former Yugoslavia seemed to have recovered from the Balkan War better than others – unfortunately, not Bosnia and Herzegovina. Driving through the small city of Mostar, the damage was evident on the scarred walls of many buildings we passed. People still occupied most, while some lay in ruins, awaiting final demolition.

With evening approaching, we made our way towards a campsite on the outskirts, intending to visit the city the next morning. Arriving at Camp Wimbledon, Valtah, an Austrian, immediately made us feel welcome, insisting we join him in having several shots of schnapps before we got settled. It was hard not to gasp after each shot of the potent liquid, and it certainly made erecting our tent a lot more challenging. We spent the rest of the evening thoroughly entertained by Valtah and a group of young Germans and Bavarians.

The next morning, after a traditional breakfast of bread and omelettes, we wandered through the maze of cobbled streets towards the old town and the infamous *Stari Most*, the Old Bridge of Mostar. Rising twenty metres above the Neretva River, the bridge was senselessly destroyed during the war and then beautifully restored since. Becoming a popular tourist attraction, local boys dove off it into the frigid water once they had collected enough money from onlookers.

We looked on in amazement as a wiry, young local took the plunge.

With our travels taking us further east, we were beginning to notice subtle changes in architecture and culture. The bridge, one of the most exemplary pieces of Islamic architecture in the Balkans, was not the only such structure we saw. Many of the houses were also designed in the Turkish style. In the city of Mostar, steeples and minarets pierced the skyline while church bells and the wailing Islamic *adhan* competed for the attention of the faithful.

After Mostar, we made our way further inland, towards the capital city of Sarajevo. We shared the road with a variety of vehicles, from slow-moving Soviet-era Trabants and Yugos to modern, luxurious limousines with official licence plates driving recklessly at breakneck speeds. The relatively empty roads tempted us to push our speed beyond the posted limits.

The locals appeared to have an uncanny ability to know where the police liked to set up their speed traps, although we never actually saw a radar gun in use. Instead, an officer would wave at us with a little red lollipop and point to the side of the road where he wanted us to pull over. On most occasions, they seemed rather baffled by what we were riding – too wide to be a motorcycle, yet not enough wheels for a car. By the time they had figured it out, we'd already passed them by, breathing a sigh of relief on our uninsured machines.

That relief didn't last long.

Sarajevo suffered more than most cities during the Balkan War. Relentlessly shelled during a three-year siege, it was still struggling to recover. Bullet holes riddled many of the buildings' facades, and pavements had red resin filling in holes from mortar shell explosions – known as the "Sarajevo Rose," marking the locations where three or more people died. The hollow remains of derelict buildings were gradually being replaced, often using brick that didn't match the rest of the building, perhaps intentionally, so that no one would forget the war.[7]

Nevertheless, the city had an optimistic air to it. The historical centre still bore many scars, but it was as lively and vibrant as any major city we had been to previously. Like Mostar, it also had that same mix of east meeting west, which added even more character. We spent a day and a night wandering through the Ottoman-era bazaars, savouring the smells and tastes of the local delicacies. The delicious grilled mince patties and stuffed pies were dangerously good. We strolled through side streets and quiet alleys, poking our noses into anything that looked interesting.

Soon, the call of the road beckoned once more.

A beautiful road took us towards Sutjeska National Park before turning south and closely following the border of Montenegro, dropping us back into Croatia, close to Dubrovnik. The Bosnian border guard gave us a thumbs-up and confidently declared, "You do have correct papers for the motorcycles," before waving us through.

It was the Croatian officials who proved to be difficult on this occasion. Instantly spotting our lack of coverage, they called a local insurance sales representative who happily drove halfway up a mountain to the border crossing to sell us a policy. I tried to persuade the border guards to let us enter without a new policy, but they were adamant about their position.

When he arrived, the cheerful, young salesman approached us with a spring in his step. "Hello, my name is Ivan. You need insurance for Croatia?"

"Maybe," I responded impatiently. We had been waiting for quite some time.

"I have just the insurance you need for my country. Please come." He motioned for us to join him inside the small border post office.

As we followed him, I said, "I explained to the border guards that we have insurance coverage for all European Union countries. Croatia is part of the EU, is it not?"

"Yes, we are now EU," he replied proudly. "But not yet for insurance purposes. I have policy for Croatia, Bosnia, Serbia, and Slovenia. Perfect for you." He smiled as he listed most of the countries we had already passed through.

"What about Montenegro? Will this new policy cover us there?"

"No," was his response.

"And what about Albania . . . or Macedonia?" I tried to hide my growing frustration.

"No, and no. Just the countries I have mentioned, no more. You must have if you want to continue."

"How much?" asked Sarah. I could tell she was resigned to the fact that we would have to pay.

He took out his phone and typed a number onto the screen for us to look at – he appeared too embarrassed to say it aloud. One of the border guards looked over my shoulder and audibly whistled when he read the amount. I did a quick conversion in my head and realised it would cost us six times our daily budget for a policy we would utilise for less than a week.

"Fuck!" I exclaimed, a word recognisable in most languages.

We had little choice but to pay it in order to proceed. I suspected we were being swindled and wondered if this was a regular arrangement between the border guards and the insurance salesman. As he filled out the paperwork, we counted out the money, realising we would have to make up for this unexpected expense elsewhere.

Things were tight enough with our budget, as it was.

Once the transaction was complete and we had our new insurance policy, we quietly returned to our bikes and began the slow descent out of the mountains. The rough road bounced us around and had us feeling a little extra tender afterwards, or maybe that was still the sting of being ripped off.

Returning to the coast, warmer weather and spectacular sunsets welcomed us as we rode towards Dubrovnik. The city exceeded all of our expectations, and we spent our time wandering through its mediaeval centre, encircled by intimidating fortress walls, on the shimmering shores of the Adriatic Sea.

On the outskirts of the city, we found the haunting remains of the Hotel Belvedere, which still stood in ruins after being attacked during the war. Supposedly off-limits to the public, it was easy to access through one of its many broken windows. Glass crunched underfoot as we wandered the deserted maze of halls, imagining how it must have looked during its heyday. Colourful graffiti adorned the walls. An open-air amphitheatre overlooked the water, providing a superb view of the old city while allowing access to many recessed rocky alcoves, ideal for a spot of afternoon skinny-dipping.

Once we had scared off most of the wildlife and several locals, we donned clothes over our damp bodies and thought a stroll in the sunshine would be the best way to dry off. We followed the shore until we reached the old fortress, where a vibrant atmosphere awaited. Bars and restaurants buzzed with the sound of hungry diners and gentle laughter. We couldn't resist the mouthwatering aromas that filled the air, so we joined the crowds and treated ourselves to a romantic meal of freshly caught grilled fish and a local wine.

It was the perfect way to end our visit.

***

From Dubrovnik, we took the coastal road to Montenegro, unsure of what to expect as our new insurance policy would become ineffectual on leaving Croatia. The Croatian border guards gave us a cursory glance and waved us on. The Montenegrin official inspected our paperwork but seemed satisfied with what we had, so we were soon enjoying one of our biggest surprises of the journey. With few expectations, we drove around the breathtaking Bay of Kotor, making repeated stops in an attempt to absorb it all.

The northern hemisphere's most southerly fjord, flanked by steeply rising limestone mountains, contrasts sharply with the turquoise waters of the Adriatic Sea. Our planned route took us up into the Lovcen National Park on what we had thought would be a fairly major road. Barely wide enough for two vehicles and practically deserted, it climbed higher than we had thought possible before levelling out on a lonely plateau dotted with old farmhouses, surrounded by stunning views of the dark peaks that gave the country its name. Black Mountain, the literal translation of Montenegro, suddenly made sense.

Without a set destination, we rode into the afternoon, steadily moving north as menacing clouds began gathering on the horizon. Anxious to find a place to camp before the weather turned ugly, we spotted a sign that said "Motorcycle Club and Camping." With dark skies looming and the prospect of steady rain for the next two days, the thought of sleeping in a tent was losing its appeal.

Still, it was better than nothing.

We turned down a smaller road that eventually turned to dirt before reaching Etno Selo Montenegro. This traditional mountain village, situated 1,400 metres above sea level, was owned and operated by the Blagojević family. After a quick tour of the property, we discovered we had better options than just camping in the rain. Smiling and nodding at each other, we opted to stay in one of their cosy little log cabins. We decided to conveniently forget about how much we just spent on insurance.

Even though the village was scheduled to close the following day for the winter, our wonderful host, Ivanka, promised us we could stay as long as we wanted. She told us that in the summer, it was often fully booked, but this late in the season, that wasn't the case at all. We had the entire place to ourselves. Other than electricity and coffee, the village produced everything it consumed, and they hoped to get completely off the grid soon. After she cooked us a hearty traditional breakfast on the first morning, the warmth and charm of this hidden paradise took hold, and we signed up for another night.

Once the weather improved, we left our delightful cabin and took the advice of a local guide who had recommended a "special road" through Durmitor National Park. As soon as we turned off the main highway, we understood why the guide could barely contain his excitement as he shared with us his knowledge of this route. Our special road spiralled up inside and outside the mountains through rough-hewn tunnels littered with fresh rock slides, clinging to the side of cliffs where no road belonged.

Soon, we were within the Durmitor Mountain range, part of the Dinaric Alps, consisting of fifteen peaks, including the highest point in the country: the 2,522-metre Bobotov Peak.[8] Here, hardy villagers were busy preparing for the coming winter. Women toiled in the fields, collecting the last harvest, while men gathered firewood or repaired their stone cottages.

Again, it looked like a different time period than the one we lived in.

While the weather had not completely cleared, the low stratus clouds remaining added to the atmosphere of this natural conservation area. With every twist and turn in the road, a new panorama greeted us. We could have spent longer within the park, but at such a high elevation, even the daytime temperatures were chilly, so we reluctantly moved on.

Dropping back into the central highlands of Montenegro, we took a detour up to the Ostrog Monastery. Perched precariously in the face of a cliff 900 metres above the valley floor, the monastery offered the weary traveller a free bed and a free meal. Even though we hadn't planned to stay, the resident monks insisted we bring our bikes inside the compound for added security, and somehow, we were soon settling into the quiet bunk rooms after enjoying a simple, delicious dinner.

When morning came, the devotees arose for worship at 5:30 a.m. We rose with them but chose to watch the sunrise rather than impose further, and by 8:00 a.m., our earliest start yet, we were back on the road. Albania lay ahead, and it provided us with an exhilarating change.

We expected to encounter conditions further along in our journey, such as challenging roads, chaotic cities, and a wild variety of old and new vehicles, yet, finding it all combined in Albania left us oddly satisfied. Somehow, everything functioned fluidly in this unique nation.

The rule of the road appeared to be that the biggest vehicle always had the right of way, and the use of turn signals was nonexistent. Four-lane highways would mysteriously become one-way lanes, and without warning, perfectly smooth asphalt would rapidly deteriorate into loose gravel or mud.

Every kilometre gained was its own reward.

As for the locals, every time we were forced to stop, it became an opportunity for them to gather around our bikes and stare. Welcomes were sincere, and enquiries about our intended journey seemed genuine and open. We wished we could have stayed longer to get to know every one of these people, but we had set our sights on Macedonia and beyond. Sadly, our goal of crossing the Sahara no later than December – and even that was considered late to avoid the "long rains" further south – pushed us onwards.

We did our best to stay ahead of the storms on this trip – and in life as well.

CHAPTER THIRTEEN

Don't Worry About Those Monkeys

Personal Diary Entry: 18th October 2013

Day 158 (of 549)

24,227 km of 62,840 km (15,142 mi of 39,047 mi)

We are opening our itinerary to the subtle suggestions offered by the people we have met and the conditions we've experienced. Unexpected challenges and obstacles steer us towards experiences we might have otherwise overlooked. These gentle nudges of fate have led us to unpredictable moments of beauty, discovery, and connection.

Once again, we had the red-tape blues. With each new frontier came the uncertainty of how best to approach the uniformed gatekeepers who practised the dark arts of border control. The wealth of conflicting information from endless sources only served to confuse, and it seldom seemed to favour the motorcycle traveller.

A fastidious official on the Macedonian side thwarted our initial attempt to leave Albania by immediately pointing out our lack of insurance on our week-old policy. It hadn't been valid for Albania either, but since we crossed at an especially busy time of day, the border guards overlooked the error.

This border guard assured us we could simply buy a new policy – our third – which would cover Macedonia and all the other countries we had already passed through, where we had no intention of returning any time soon. Our stay in Albania had been so cheap we felt we could afford the added expense, so we agreed to become Europe's most-insured motorcycle riders and asked where we could make our donation to the insurance company's CEO retirement fund.

"Not here," was the response.

It turned out that the only office capable of issuing an acceptable policy was over a one-hour drive back through Albania, around the scenic Lake Ohrid, at another border crossing. We had no choice but to return through no-man's-land to the dusty Albanian post where the guard had promptly stamped us out less than an hour earlier.

When the Albanians enquired as to why we had decided to return so soon, we attempted to convince them that we had enjoyed their country so much that we decided to spend more time there, and what little we had seen of the Macedonian side just hadn't looked that impressive. They eyed us suspiciously as they rechecked our documents and soon discovered the real reason: our lack of insurance for Macedonia – and Albania.

"You have problem" was becoming a familiar phrase as we fumbled our way through our journey, especially when officialdom was involved. We were stuck between the borders of two countries, without a valid insurance policy for either one. The guard pointed towards a small, rusting shack nearby and said we must buy a policy in order to drive through Albania to the next crossing, where we would have to buy yet another policy for Macedonia. As we headed towards the shack, the border guard picked up a phone, and we soon heard an electronic warble coming from inside the insurance office.

A French rider we'd met months earlier in Germany told us of a strategy he employed when dealing with authority figures: "Smile, and talk constantly." So, the two of us went to work on the poor insurance salesman, and we soon had him convinced that there had been a terrible mix-up and that we would no longer require his services at this time.

We bade him a hasty farewell and sped off into Albania to attempt another crossing.

Every hour we rode in Albania took approximately two hours off the lives of our motorcycles, and after a bone-jarring race around Lake Ohrid, we went through the same old routine one more time. The Albanians were most agreeable to letting us leave, but the Macedonians insisted we visit a nearby booth and add their country to our ever-growing list of insurance coverage.

That was where we encountered our first taste of Macedonian hospitality.

Biljana, our extremely friendly insurance salesperson, insisted on easing our pain by inviting us inside her cramped, cosy office while she prepared some strong local coffee. While the border guards demanded we return immediately with the necessary paperwork, she turned the tables and had them keep a watchful eye on our bikes while we caught up on all the latest gossip from this lonely border post.

"You are such a beautiful couple!" she exclaimed, offering each of us a firm handshake and beaming smile. I looked over my shoulder towards our bikes and noticed a border guard watching us, shaking his head and throwing up his hands.

"Oh, don't worry about those monkeys," she continued as she rearranged the furniture to make room for us around her cluttered desk. "Now, before we do anything, you must try some authentic Macedonian coffee."

As she prepared our drinks, a guard yelled from outside. I could sense the urgency in his voice even if I couldn't understand the language. Biljana shouted back and began to chuckle to herself.

"Ah, Boris," she said with a dismissive wave of the hand, then turned to us. "He's such a hothead. What's the rush? I never get any visitors. Now, you just relax and pay no attention. Sugar?"

She poured the darkest coffee I've ever seen into three small cups before adding a mound of sugar to each. It tasted bitter and sweet and surprisingly good.

"Welcome to my country," she said, looking from me to Sarah and back. "I know you have just arrived, but you are going to love it here. It is the one and only Macedonia. Don't let those Greeks tell you any different."

For the next hour, Biljana delighted us with stories about her country, childhood, family, and friends. Whenever a border guard asked about our progress in obtaining a valid insurance policy, she would scold them fiercely and giggle uncontrollably afterwards.

It made for a very entertaining afternoon.

Eventually, we entered Macedonia with our new insurance policies in hand. We abandoned our plans to go further into the country. We settled for the quiet village of Peshtani on the shores of Lake Ohrid, almost directly opposite the camp we had stayed at the night before in Albania. We spent a restful night in a cheap apartment overlooking the lake, and early the following day, we were back on the road, climbing into the majestic Galičica National Park.

As the road swept higher, it afforded us beautiful views of the lake below and Albania beyond. We closely inspected our maps, hoping to find more interesting routes. However, it proved fruitless as few of the passes had established roads making it all the way over.

We spent a night outside the capital, Skopje, before exploring the city the following morning. Situated on the banks of the Vardar River, Skopje has an intriguing mix of modern and ancient architecture. A disastrous earthquake in 1963 destroyed eighty percent of the city, and reconstruction was still underway.[9] The 1,400-year-old Kale Fortress overlooked the city centre, where bazaars from the Ottoman Empire abutted modern plazas. On a distant mountaintop, clearly visible from the city, stood a 66-metre cross – one of the world's largest crosses – built in 2002 to celebrate 2,000 years of Christianity.[10]

The pressure to keep moving soon had us back on our bikes and moving east towards Bulgaria, with one last stop before leaving the country. Local currencies were often difficult to exchange once we had crossed the border, so we turned every penny we had into fuel at the last petrol station in Kriva Palanka, where we met Irene. A well-travelled local, she helped me complete a challenge I had undertaken along the way.

"Wow," said the young woman who approached our bikes, noting the California licence plates. "You have travelled such a long way."

"Not as far as it looks," I informed her. "We shipped the bikes from the USA to Ireland – that's where we began this journey."

"That's still a long way," she mused, then paused and stuck out her hand for a shake. "Hi, my name is Irene – like the hurricane." She laughed. "I was living in the USA when the big storm hit New York, and after that, no one ever forgot my name."

"Hello, Irene, a pleasure to meet you. This is Sarah, and they call me 'Irish.'"

After talking with Irene for a short time, I decided to ask her for a huge favour.

"Irene, I have a little problem, and I was wondering if you could help me solve it," I began. "Ever since we started this journey, I've been sending home two postcards from every country we've been to – one to my father and the other to my young nephew. On each one, I write a greeting in the local language while including some information on the country they are from."

"Oh, that's such a beautiful thing to do!" said Irene, clapping her hands with joy. "What can I do to help?"

"Well, it's Sunday, and we failed to pick up any local postage stamps and all the post offices appear to be closed for the weekend. We plan to leave today, and this will be the first country where we haven't been able to send our cards. I'd be grateful if you could take these cards and send them to Ireland for me." I showed her the postcards I had chosen and offered her what local currency I had left.

"I would be honoured. I promise I will post them first thing on Monday morning." She seemed noticeably touched that I had chosen to trust her with this task. "Leave it to me. Hurricane Irene will not let you down!"

Good to her word, both postcards arrived at their destinations several weeks later. From the warm welcome we received upon entering

Macedonia to our final departure, everyone we encountered along the way exceeded all of our expectations.

***

We had been excited about entering Bulgaria for some time as it was home to "MotoCamp," a sanctuary for travelling bikers in the tiny village of Idilevo. It provided a place to rest and work on bikes, all while sharing common dreams with people who didn't think we were crazy. We gave the capital city of Sofia little more than a quick drive through, stopping only to pick up a few spare parts from the local BMW dealership before spending the night in a cheap hostel near the centre.

The following afternoon, as we pulled into the renovated collection of old farm buildings that make up MotoCamp Bulgaria, we immediately felt at home. Even though it was late in the season, several other overland adventure riders were in residence, preparing their machines for onward travel while resisting making it a permanent base on account of the overwhelming hospitality of the owners, Polly and Ivo.

Parts took a long time to ship to this sleepy little village, but nobody seemed to mind. Michael and Tamara, an inspiring couple from Germany, had been waiting six weeks for their clutch slave cylinder to arrive. In the meantime, they had bought a house in the village.

Igor and Penny, an intrepid pair of New Zealanders riding from South Korea to the Netherlands, were rebuilding their drive shaft, waiting for bearings to arrive, when the infinitely generous Shlomi from Israel, our instant friend from a rainy day back in Slovenia, arrived carrying the exact bearings they required.

Two more bikers arrived as we settled in to take care of a few of our minor maintenance issues. Alex was returning from Russia after a harrowing encounter with their health care system, and the infectiously cheerful Alpy from Turkey was on his way to Sofia to pick up a new set of custom-made luggage bags for his bike.

The village was home to several expat bikers who dropped by regularly to see what the wind had blown in and to find an excuse for a beer at the rustic bar above the old barn. We took advantage of the great weather and local knowledge, joining a group who were test-riding their bikes before setting off to pastures new.

A vast network of twisting, mountainous roads surrounded the village, revealing glimpses into Bulgaria's recent history. Abandoned monuments honouring its communist past lay derelict on mountaintops, the most bizarre of which was the *Buzludzha* building, a relic from a period most would rather forget. The group joined us on their bikes as we went to explore.

It had looked surreal from a distance, its profile resembling a stranded flying saucer with an enormous tail-fin. On closer inspection, its scale was overwhelming. We gained access to the Buzludzha structure by squeezing through a broken window, and once inside, the grandeur of what it must have been like was still evident in the vandalised mosaics and murals. As the wind howled through the glassless window frames and water dripped from the damaged roof, the vacant building had a haunted feel. We spent several hours clambering over broken glass and rubble, exploring every nook and cranny. A spindly finger of concrete towered 70 metres over the rim of the disc-shaped base.

Leaving Sarah with Shlomi and the rest of the group, I found my way into the tower's hollow base. A rickety ladder disappeared into the darkness above me. I questioned the wisdom of trusting it with my body weight, but I couldn't resist the temptation to see where it led. I wrapped my fingers around the thin, rusting rungs and began to climb. One step at a time, ascending into the fading light, I was soon engulfed by a damp, cold blackness. I couldn't see what the ladder was tethered to, and occasionally it would tremble precariously, causing my heart to race.

I regretted not telling anyone where I was and what I was doing.

I was sure I must be close to the top, but the ladder kept climbing, and so did I, too stubborn to admit defeat. Finally, a faint red light appeared overhead, and I wondered if some part of this building was still alive. After what seemed like an age, the ladder terminated at a thin metal mezzanine. Glass crunched beneath my heavy boots as I pulled myself up with my hands firmly gripping the flimsy safety railing. I was inside the enormous red star that had been the beacon for this curious edifice. A chilling wind howled through the ragged holes in the tinted glass.

"You idiot," I whispered, gazing into the black abyss below before descending into the gloom.

***

Returning to MotoCamp, Sarah and I got to work on our bikes, breaking them down and changing out a few essentials. I had to attend to a stripped sump plug that was turning a regular oil change into a two-hour ordeal of removing the sump pan and replacing the gaskets.

It seemed as though BMW had thought it appropriate to make the drain plug out of a grade of aluminium with the same consistency as butter. After removing the sump pan and attempting to capture all the used oil in the messy procedure, I timidly encouraged the plug to unscrew with a hammer and punch, fearful of damaging the brittle aluminium pan.

Convinced the nearby town of Sevlievo would have a mechanic with a bolt extractor, I took the pan in to seek their advice. After a quick round of charades, I got the idea across of what I wanted, and they got immediate results by using a bigger hammer and a bigger punch.

With a sheepish grin and a bruised ego, I thanked them for their help and raced back to camp to complete the service. We reassembled our bikes, performed a few tests, and finished the day with a cold beer on the lawn as the sun sank below the horizon. All too soon, our brief stay at MotoCamp Bulgaria had come to an end, and it was time to bid farewell to our new friends.

We had lost count of the number of goodbyes we had said along the way. One of the harder parts of this type of travel was as soon as we'd made a new friend, it was time to leave, and with our departure came an uncertainty of whether we would ever see them again.

We met some incredible people along the way, other travellers on adventures of their own. Some of them were on foot or bicycles, some on motorcycles or cars, and some even on boats – and it felt reassuring to exchange stories and encouragement.

But there were others who would hope to convince us that our trip was foolhardy and irresponsible. On some occasions, we even questioned our motivations, especially when plans went pear-shaped. To those who had helped us along the way with kind words and good advice, we would be forever grateful.

Their well-wishes became our tailwind, nudging us on whenever we faltered.

CHAPTER FOURTEEN

Surrendering to the Lessons

Personal Diary Entry: 30th October 2013

Day 170 (of 549)

26,158 km of 62,840 km (16,349 mi of 39,047 mi)

Our journey has been influenced by circumstances beyond our control and a new path awaits us, like a ribbon of promise unfurling before us. The road is carrying us towards ancient lands, where myth and history are woven together in a rich fabric of wonder and intrigue.

In the early stages of planning this trip, Sarah and I had assumed between one to two years would be ample time for our adventure, but now we realised it would barely be enough.

We turned south towards Turkey, stopping for one last night in Bulgaria close to the border, determined to approach the new frontier after a good night's rest.

Leaving Bulgaria the following day was hassle-free, but entering Turkey involved visiting multiple checkpoints. Before long, they were demanding their pound of flesh, visa fees, customs clearance, and, unsurprisingly, a new insurance policy – our fourth. Just when we thought we were on our way, a wrong turn took us onto a toll road where we were required to purchase a motorway pass for each bike, even though we exited the highway at the next available opportunity.

After spending a couple of hours being driven doolally attempting to get permission to cross into Turkey, we decided to avoid the bottleneck of Istanbul and turned south towards the Dardanelles, riding along the Bosphorus Strait. Exhausted, we decompressed by spending the night at a cheap hotel just outside of Sarkoy.

The following day, we pensively rode through the Gallipoli region, surrounded by numerous memorials celebrating a victory or commemorating the fallen. These were the solemn reminders of the human cost of lines on a map. The same lines each country protected were the ones we had difficulty crossing.

By midday, we were boarding the ferry at Kilitbahir, which would take us into Asia by way of Çanakkale and onto the Turkish mainland. By now, the scale of the country was becoming apparent. We'd covered half the distance we had intended to and still had much more to go. Following the Aegean Coast, we passed by the ancient ruins of Troy and Xanthos, resisting the urge to spend more time in the area.

"Another time," we said in these moments. We instead kept moving, reassuring each other that there would be plenty more opportunities to travel together.

At that moment, we had rubber to burn. The quality of the roads matched anything we had ridden so far, and the kilometres rolled by as we fell into a gentle rhythm of leaning left, then right. Distant Greek islands appeared to float on a delicate bed of haze under deep blue skies as our route twisted south along the water's edge. With night falling and the temperatures plummeting, we pulled off the road and chose a rough, wild camp, where we spent the night shivering inside our down sleeping bags.

After an early start, we spent several days motoring our way across the high, barren plains of central Turkey before reaching our goal – the town of Göreme, in the Cappadocia National Park. Getting off our bikes and hiking after spending so many hours in the saddle felt good.

The region's soft, volcanic rock, in a variety of pastel shades, had been weathered into hundreds of wind-carved columns. The local villagers had then chiselled homes and churches into these towers to create an alien landscape perfect for the curious visitor. From sunrise to sunset, we wandered among the dreamy fairy chimneys towering above us. Every opening at the base of a stone pillar provided an irresistible invitation for further investigation, and a thin layer of light dust soon covered us.

Sinister events in Cairo disrupted our plans for catching the ferry from Turkey directly into Egypt. Riots were being met with fierce oppression as the Arab Spring destabilised the status quo in Egypt. News of intense fighting in Syria and a border on full lockdown made the overland route through the Middle East impossible. We looked into other options and finally resolved to turn west into Greece, where we could catch a ferry to Israel and attempt to enter Africa via the Sinai Peninsula.

Slowly, it dawned on us that any attachment to a fixed agenda could only lead to frustration and disappointment.

At times, we felt like mere passengers on this grand adventure as we loosened our grip on control and surrendered to the lessons the journey had to teach us about ourselves and our odyssey. Embracing the unfolding serendipity, we discovered a feeling of freedom and liberation.

Releasing the burden of expectations rewarded us with an oddly satisfying sense of relief and excitement. We opened ourselves to the unexpected, finding beauty in the detours and dead ends. The journey became about more than just reaching the next destination – we appreciated the transformation within us along the way.

We discovered our resilience in the face of uncertainty, adaptability in moments of change, and ability to find joy in the simplest of encounters. We grew as individuals, approaching the unknown with a newfound sense of curiosity and wonder.

The road became our guide and providence, our compass.

***

Leaving Göreme proved difficult. We'd found a quiet campsite overlooking the surreal town, perfect for capturing the spectacular sunrises. Parked next to us were fellow travellers from California,

Scotland, and Ireland, who all offered good advice and welcome company. Our travels through Turkey had left us wanting for more, but a storm was coming, and as usual, we hoped to outrun it. So, amid light rain and strong winds, we packed our tent and made preparations to move on.

It took several demanding days to traverse this enormous country.

We weren't successful in our attempts to avoid the foul weather and got caught in the middle of it once again. We rode with determination through heavy winds and bitterly cold rains.

Exhausted at the end of such an arduous journey, the prospect of navigating the daunting capital with its chaotic traffic held little appeal. We decided to abandon our plans to make it to Istanbul and pulled over a few hours beforehand at the resort town of Akçakoca on the shores of the Black Sea.

Wearily, we searched for a place to rest and recuperate. After pricing several hotels, we opted for a much more affordable *pension*, a small room in an apartment building overlooking the waterfront, from where we could easily see our bikes parked in the street below. We unloaded our luggage inside and decided to take an evening stroll before retiring for the day.

As darkness fell, we walked along the boardwalk near huge waves pounding the shoreline.

"So Syria is out, and now the ferry to Egypt has been cancelled. Does it feel like the universe is telling us something?" I pondered aloud.

"Sure feels that way," Sarah replied. "I hope nothing goes wrong with the Greece-to-Israel plan."

"Only time will tell, I guess."

I tried to think of viable alternatives, but it hadn't been easy getting up-to-date information on conditions throughout the Middle East. "You know, we could do something else . . . "

Sarah looked up at me inquisitively with her beautiful, hazel eyes, waiting for the rest.

"We could just abandon the Africa route and continue east through the Stans and into Mongolia," I suggested.

"What?" Sarah sounded surprised. "You mean make it a 'round-the-world' ride instead?"

"It's just a thought I'm throwing out there in case we can't get into Africa," I added. "It might be good to have a plan B."

"I think we already burned through our plan B," she said with a dry laugh. "We must be on E or F already."

Sarah stopped on the beach and turned towards the stormy Black Sea. The wind pushed her hair from her face and her skin glistened from the light mist thrown up by the crashing waves. She looked captivating. I watched her contemplate for a while in silence. Then she sighed and turned towards me. "I think we should continue and at least give it our best shot. Africa sounds so much more . . . enticing."

"Okay," I welcomed her reassurance. "Africa it is then. Let's make it happen."

"Hey, maybe we'll get a chance to see Shlomi again. He promised he would show us a good time if we ever came to Israel." She smiled at me and took my hand.

***

Our new friend, Alpy from Turkey, who we met at MotoCamp in Bulgaria, had invited us to stay if we ever passed through Istanbul. As blue skies broke through the dark clouds, we approached the outskirts of the sprawling metropolis, fully rested and thrilled by the challenge of taking on its infamous congestion.

Long before we expected to enter Istanbul, we were engulfed by Turkey's most populous city. With an estimated fourteen million inhabitants, its scale was truly impressive.[11] Sticking to the elevated highways where traffic was most predictable, we pushed further into the heart until we had no choice but to enter the city streets.

Suddenly, the rhythm of the road changed dramatically as cars, buses, heavy trucks, and motorcycles all competed for the limited opportunity to move ahead in the traffic. Initially overwhelmed, we quickly became accustomed to the chaos and joined in the fun.

Our motorcycles made the going a little easier than it was for those in cars, but, loaded as they were, we couldn't keep up with the local bikers. We followed our rough directions, reading foreign signs while hoping to anticipate the next assault in our immediate vicinity. Other motorists seemed oblivious to our presence. We still made good progress and, before we knew it, a warm smile and a waving hand beckoned to us from the footpath.

We had made it to Alpy's.

A seasoned motorcycle traveller himself, Alpy quickly made us feel at home, encouraging us to pull our bikes into his covered garage alongside his own. We stripped off a few essentials and followed him up to his apartment.

Our introduction to Istanbul began with an evening cycle around the Kadiköy district on the Asian side of the city. A lively, bustling centre of tea houses, street stalls, and food courts, it was surprisingly busy despite the late hour. We dined on *lahmacun,* a kind of Turkish pizza, and a refreshing salted yoghurt drink known as *ayran,* while we watched the locals go by.

The following morning, we caught the ferry into the European side of the city, intending to explore the historical centre. On the way to Taksim Square, the site of the recent anti-government protests, we stumbled into the hardware district, where we spent several hours sourcing various additions for Sarah's motorcycle.

The city's older commercial districts appeared to have become rather specialised in what they sold, and we discovered many streets dedicated to one particular item. One street sold nothing but nuts and bolts, while another was entirely dedicated to buttons, and so it went. With a combination of our very limited Turkish and some simple drawings, we found everything we needed within walking distance.

We spent the rest of the day wandering through the Grand Bazaar, the world's first indoor mall, and around the enormous mosques that dot the old city. Established by the Greeks as Byzantium in 660 BC, this transcontinental metropolis had served as the capital of the Roman,

Byzantine, Latin, and Ottoman empires, each leaving its indelible mark upon the city's architectural heritage.[12] Modern tower blocks shared the skyline with needle-like minarets that pierced the horizon in a seemingly endless panorama.

On our third day in the city, we decided to track down a mechanic with a chain breaker so we could replace the tired chain and sprockets on Sarah's bike. We found more than we were looking for within a few streets of Alpy's apartment. A bike-cleaning service adjoined a friendly mechanic's workshop, all next door to a cheap hairdresser's salon. While we cleaned our filthy motorcycles, we met many local bikers who offered good advice and overwhelming hospitality, one of whom kindly insisted on taking us out for a delicious lunch. We gratefully took him up on the offer.

Afterwards, with clean machines, we felt much better about approaching the mechanic who dropped everything and immediately set to work on Sarah's bike. Meanwhile, he allowed me to use his tools to make some alterations to her luggage system. Sarah even ventured inside the salon and had her hair cut by the resident hairstylist.

Overall, it turned out to be a most productive day and all at a reasonable cost. We often preferred to fumble our way through our maintenance procedures, but it was nice to occasionally pay a local expert so we could avoid some of the frustrations. With the bikes looking and performing much better, we returned to Alpy's and made arrangements to leave the following day.

After another inevitable farewell early the next morning, we were back on the road again. We had hoped to avoid the worst of Istanbul's rush-hour traffic, but in a city so huge, every hour of the day was like its busiest. Even though we thought we were moving quickly, it still took almost two hours to clear the western edges of the city. We avoided the major highways and followed the coast along the Sea of Marmara before turning inland to the border with Greece.

By midday, we noticed an escalation of military activity alongside the road. There was still some tension between Turkey and Greece, but it had eased considerably in recent years. When we arrived at the Turkish checkpoint, I handed my documents to the border guard on duty. His furrowed brow indicated we could have a problem.

Although we were still uncertain why, it appeared my bike had been incorrectly logged into their system when we had entered the country. My licence plate contained the letter "O" twice, which could have been confused with a zero. It took the first border official several attempts to enter my details before he called his superior. Meanwhile, a long line of traffic began to form behind us.

In the end, it took four border officials, who consumed many cups of tea while smoking multiple cigarettes, over half an hour to enter all possible combinations of my licence plate into their computer before they allowed us through. We waved, apologetically, to the patient drivers in the line behind us before opening our throttles and powering into Greek territory. Well-armed soldiers from both sides wearily watched our bikes as we slipped across yet another invisible boundary. We hoped the Greek border guards wouldn't notice Sarah was really stretching her welcome in Schengen territory. Thankfully, they gave us little more than a cursory glance before waving us through.

We were immediately impressed by the Greek roads.

As we rode through the first few kilometres of the heavily militarised border zone, the smooth surfaces and clear road markings came as a pleasant surprise. Our delight lasted approximately thirty kilometres before the surface deteriorated and the road markings vanished.

With autumn well underway, the eastern region of Greece felt deserted. The summer crowds had long since departed, and we had the roads almost entirely to ourselves. Empty camps allowed us to sleep to the sound of gentle waves rolling on abandoned beaches.

We continued our journey around the coast through small towns where residents prepared for a quiet winter without tourists. Soon, our road turned north and wound its way back towards the mainland and into Thessaloniki, where arrangements were underway for the upcoming festival of Demetrius. Traffic police lined the streets and barricades were being put in place to cope with the large crowds arriving to take part in the celebrations.

We stopped briefly in the city for lunch before continuing along the coast towards Mount Olympus National Park, searching for a campsite to use as a base for a few days while we attempted to climb Greece's highest peak. To our dismay, all the campgrounds we found were clearly closed for the winter. Fortunately, we discovered a small guest house on the waterfront in the tiny village of Litochoro. Summit Zero Hostel proved to be a lucky find. Pericles, the owner, was an avid mountain runner who gave us plenty of advice on routes up to the summit.

Viewed from the shoreline, the impressive silhouette of Mount Olympus would occasionally break through the clouds, and as the sun set behind it, an ominous shadow spread over the quiet village. The tranquil hostel had only a few guests at this time of year, so we had room to roll out our gear and take care of a few outstanding chores.

We rose early on our second day to drive to the base of the mountain.

Pericles had warned we would be in for a long, gruelling day, so we left before sunrise, quietly slipping out of the hostel and into the early morning haze. A winding road took us partway up the eastern side of the mountain's flank, and we soon emerged from the mist into a spectacular sunny morning as the sun breached the blanket of fog that lay over the ocean below. The orange glow lit up the fluffy white undulations, setting the horizon on fire.

It was still chilly when we reached the trailhead, so we wasted little time securing our bikes and donning our hiking gear. We had gained a little by driving to the trailhead, but with the summit at 2,917 metres (9,570 feet), we knew we would be in for an arduous ascent. With multiple routes to the summit, we had agreed beforehand to hike separately and meet at the top. It was time to spend another day apart. We kissed and bade each other farewell before splitting up.

Months in the saddle had taken its toll on our overall fitness, and we were feeling it as we reconnected at the top six hours later, hearts pounding and legs burning. The effort paid off – the views were astounding and the weather perfect. We spent a long time on the peak, enjoying a well-deserved break and a light lunch before descending together.

The last few hundred metres before the top had involved a rock scramble, which felt significantly more exposed on the way down. Multiple plaques were attached to the rock face with words in Greek and dates commemorating less fortunate climbers. Returning to the hostel, invigorated but exhausted, we had an early night.

From the coast we turned inland, having chosen a more intrepid route towards the region of Meteora. Our road twisted south around Olympus and onto the plateau behind, gradually becoming smaller and smaller. Before long, we were driving on dirt roads into deep, golden forests, the surface coated with colourful autumn leaves.

When we eventually broke through the tree line, our road deteriorated further and soon we had to pull over and reassess our plan. Heavily rutted and barely visible in places, the road didn't give any indication as to where it would connect with a more viable route or whether it would vanish altogether.

We decided to err on the side of caution.

With limited fuel and unreliable maps, it seemed like the smart thing to do, so we turned back and found a slightly better road. Even though our two-hour drive became a five-hour adventure, we finished the day unscathed, smiling as we relaxed over cold beers.

CHAPTER FIFTEEN

Delaying the Inevitable

Personal Diary Entry: 11th November 2013

Day 182 (of 549)

26,115 km of 62,840 km (16,322 mi of 39,047 mi)

A growing sense of unease is gnawing at the edge of my mind. I try to rationalise this prickling anxiety as nothing more than an overactive imagination. Yet, the knot of apprehension that tightens within my gut seems to herald the approach of a different kind of storm, something inevitable and inescapable.

The peaceful village of Kastraki, in Central Greece, became home for several days while we explored the wondrous monasteries of Meteora. These monasteries were perched precariously on top of enormous rocky pinnacles, the contorted shapes of the ancient rock formations supporting the more familiar dimensions of architectural designs hundreds of metres above the valley floor.

It was hard to imagine how anyone could build such extensive structures atop these steep-sided pillars of stone.

Only six remained in use, each accessed by a modern rope bridge, cable car, or a tunnel carved up through the rock. Connecting the bases of all six sites was a wonderfully smooth, twisting road from the village below, too beautiful to resist driving multiple times.

With the prospect of acquiring yet another fresh set of tyres in the city of Patras further south, we picked out an interesting route through the mountains which would take us into the Peloponnese region. That turned out to be another highlight of our trip: empty roads occasionally cluttered by recent rock falls, with breathtaking vistas of the mountainous terrain. Silky, sweeping turns eventually led down to the pristine coast, making for an exciting ride.

The beaches of Greece were stunning, boasting fine white sand with crystalline waters. It was almost a shame that we'd ordered the tyres to pick up further along our route, as we kept wanting to stop and savour the sunshine at each beach.

The friendly staff at a motorcycle store in Patras had agreed to supply us with fresh rubber and a few extras. We pulled up to their shop early on a Friday and set about fitting a new set of Heidenau K60 Scouts to my bike while the whereabouts of Sarah's tyres remained a mystery. Numerous attempts to locate them proved fruitless. Although our current tyres still had a little life left in them and had performed well beyond our expectations, we felt as though a more aggressive tread pattern would be better suited for the roads we were likely to encounter in Africa.

Sarah also thought a set of handlebar risers would enable her to ride more comfortably in the standing position. More challenging terrain required a different style of riding. Shifting weight onto the feet and moving one's centre of gravity forwards would allow the legs to absorb more aggressive undulations. Raising the height of the handlebars helped ease some of the resultant strain on the lower back, so adding a spacer between the front forks and the bars made for more comfortable ergonomics.

While Giannis and his team at MotorAid tried to pin down the exact location of Sarah's new tyres, we pulled out our tools and set to work on her bike, fitting the new risers. By late afternoon, it was apparent that her tyres were on their own tour of Greece and would not arrive until after the weekend, so we set off to explore the Peloponnese area, ending the day at a quiet campground on the coast.

The following day, we rose early before moving further south, through Sparta to Mystras, spending the night at a beautiful hotel in the shadow of the Taygetos Mountains. Surrounded by ruined palaces, temples, and monasteries, the ancient town of Mystras once served as the effective capital of the Byzantine Empire.

It was unusual for us to have a place to be or a set schedule, but the arrival of Sarah's new tyres had us returning to Patras before we could fully appreciate the area. We had them promptly fitted and were soon back on the road. Crossing the Gulf of Patras for a second time, we took the road east to Athens.

With our route taking us through Delphi, we felt obliged to stop and consult with the Oracle about our decision to undertake this odyssey. In ancient times, Delphi was considered to be the centre of the known world and home to Pythia, the high priestess of the Temple of Apollo. Also known as the Oracle of Delphi, her counsel was highly regarded and she was once thought to be the most powerful woman of the classical world.

Sadly, she appeared to no longer be in residence, having retired centuries before. Or perhaps she could sense my Celtic heritage. After all, it could have been my ancestors who raided her temple, plundered the sanctuary, and stole the unquenchable fire from her altar.

Hell hath no fury . . .

Dark clouds accompanied us on our descent into Athens, and as we arrived at our camp, we could see we had just missed the worst of the weather. Sleeping bags and clothing hung from every available hook as pools of fresh rainwater surrounded bedraggled tents. We chose our site carefully, avoiding any obvious low spots in case more heavy rain was forecasted. The weather gradually improved over a couple of days.

We stripped my bike of its luggage and rode two-up on one bike through the city, exploring Athens, sourcing the last of our spare parts, before our imminent exit from Europe. We carefully observed the local bikers to see what was considered acceptable riding etiquette. Splitting traffic, nimbly accelerating between slower-moving vehicles, was expected. At each red stop light, a wave of motorcycles filtered its way to the front of the line so that each time the light turned green it felt like the starting grid of a motorcycle race. Engines screamed and rubber squealed as we raced to our destination: the next red light.

My powerful engine gave me a slight advantage; however, I lacked bravado even with our horsepower advantage as the more seasoned city riders barely squeezed through the minuscule gaps that briefly opened between cars and buses. By the end of our first day in the bustling city, I was beginning to get a feel for the more aggressive riding style.

As we kept to the inner confines of the busy centre, we noticed a strong police presence. Fortunately, they seemed to show little interest in my riding as I defied one-way streets and performed illegal U-turns with the local bikers.

Exploring the sprawling city of Athens became less daunting as we caught occasional glimpses of its most recognisable landmark – the Acropolis. Overlooking the city, it served as a helpful beacon for navigating the labyrinth of city streets below. Talking to each other over our radios while I manoeuvred around automobiles and motorcycles, we made a plan to visit the monument the next day.

On our first day, we took care of business.

We visited multiple motorcycle shops, gathering most of what we needed. The crew at BMW Athens were particularly helpful and even gave us a significant discount when they learned of our plans. We received a warm welcome at the Touratech store: the largest supplier of adventure motorcycle parts and accessories and an international motorcycle travel community favourite. Even though we didn't buy anything, the friendly staff wouldn't let us leave without taking a few freebies.

On our second day in the city, we began to enjoy the city's liveliness, manoeuvring through traffic more expertly as we headed towards the Acropolis. We hopped off the bike and hiked up to the summit, past the exquisitely restored Theatre of Dionysus. Once on top, the true scale of Athens became apparent as it spread out in every direction below us.

The Acropolis is arguably one of the most important ancient monuments in the Western world. But the Parthenon, its most prominent structure, draws the most attention. Completed in 438 BC, its designers manipulated its size, scale, and form so that, to the eye, it appears perfectly balanced.[13] Each column was subtly crafted with a slight swelling at the midpoint for the effect of appearing straight when viewed from a distance. Every dimension was based upon one rudimentary measurement taken from the human form. For such an imposing structure, it still yielded a sense of timeless grace and harmony.

With our impending departure from Greece now looming, we returned to camp. We packed our gear with the prospect of spending our last night in Athens on the southern side of the city with Dimitris, the local Horizons Unlimited representative. Once again, the adventure traveller's online forum served us well.

Meeting Dimitris was like being reunited with a long-lost friend.

With overwhelming hospitality, he invited us into his home to get settled before taking us out on an off-road adventure in his compact two-wheel-drive Mazda. We made it to a local mountaintop just in time to witness a spectacular sunset as we gazed over the city and the nearby Marathon Bay. Full of knowledge, he proudly pointed out many of the local landmarks, enlightening us with glimpses into the region's remarkable history.

As night fell, we returned to the city, where he insisted on treating us to a delicious feast at a local Cretan restaurant. After numerous shots of a mysterious local beverage, it was well into the wee hours of the morning before we returned to his apartment, dreading the early morning wake-up call and the race to the southern port of Lavrio.

After a comfortable but short night's rest, we woke with furry mouths and cloudy heads, said our goodbyes, and climbed onto our bikes before speeding south for our 9 a.m. rendezvous with Salamis Shipping.

Our over-enthusiasm to be on time saw us at our shipping agent's office long before we needed to be. We began the slow process of having our documents and bikes inspected while waiting for the boat to arrive. As the day wore on, we completed all the necessary paperwork and readied ourselves to bid Europe farewell – many times. Fully prepared, we spent the day at port, scanning the empty horizon for anything resembling a cargo vessel.

We finally caught a glimpse of an approaching container ship at dusk, and it was fully dark by the time we rode on board. After waiting for thirteen hours, a change in engine's pitch heralded our departure, and the next leg of our journey began as we sailed into the night.

***

After our slow voyage across the Mediterranean onboard the *Alios* container ship, we arrived early in Haifa, Israel, well-rested and eager to get back on our bikes. After the boat had docked, a group of security officials came onboard to begin the slow process of clearing us through customs and immigration. When they finally established our intentions, we were allowed to take our bikes off the boat so they could more thoroughly inspect them.

Proof of entry into Israel can create problems when applying for visas to enter other countries. Because of that, an immigration officer met us as we cleared the loading ramp and issued a special entry stamp attached separately to our passports so it could be removed on exit.

We continued the security checks on land, removing articles from the bikes to be x-rayed and inspected. As soon as we had satisfied the security personnel that we were harmless, they led us to a bonded parking area where we had to store the bikes until we completed the rest of our paperwork. The bikes needed to be insured to obtain a temporary importation permit and drive out of there. They allowed us to exit the port on foot to visit a nearby insurance broker.

Our good friend Shlomi had agreed to meet us at the port. We didn't realise that this bureaucratic runaround would prevent us from leaving with him until the very end of the day.

With what we thought was the necessary documentation, we returned to the bonded parking area and asked for our bikes, only to be informed that we still had more to do. Unfortunately, the parking area and the clearance offices were some distance from each other, so each time we had to walk from one to the other, it added another half hour to the whole process. As the day wore on and we jumped through the required hoops, we noticed it was getting dark. We suspected we would not get our bikes cleared until the following day.

We readied ourselves for receiving the bad news.

Exhausted and ready for it all to be over, we were grateful when we received good news instead. The last person we had to satisfy was willing to stay late just to release the bikes. As we made our way to the final security checkpoint, we high-fived and almost skipped like happy little children back to the parking area.

At the final checkpoint, an impossibly young guard noticed a missing digit on the vehicle identification number on Sarah's importation papers. He informed her she would have to turn back and have it rectified. Surprisingly, the customs office was still open by the time we returned out of breath after a brisk run along the path we'd taken many times before. After they resolved the problem, we were finally allowed to leave.

By now, it was nighttime, and we had been at the port since dawn. The entire time, Shlomi was waiting patiently outside. Without access to a phone, we had struggled to keep him informed of our situation.

He didn't seem to mind at all.

After bear hugs all around, the three of us hopped on our bikes and rode out of Haifa. Briefly stopping in Tel Aviv to snack on some tasty falafel, we embraced again as we waited in line, overjoyed with our luck to see each other again so soon. By the time we reached the Negev desert, our destination for the evening, it was 2 a.m. Following a few celebratory shots of the local beverage, *arak*, it was off to bed. Sarah and I slept well into the next morning before beginning our first excursion into the region.

Over the next two weeks, using the tiny settlement of Ashalim as a base, we explored much of the diverse country; from the fertile forests of the north, to the barren deserts of the south; from the Dead Sea in the east to the golden beaches of Tel Aviv in the west. Shlomi rode alongside us on his BMW R 1150 GS as our guide to some of the less-visited parts of his country.

An invitation to join a group of local riders enabled us to ride along the normally closed Sinai border road, a deserted stretch of twisting highway bordering Egypt. We wild-camped on the shores of the Sea of Galilee and crossed the Jordan River. Over the following days, we wandered the ancient markets of Jerusalem and had our efforts to swim in the Dead Sea thwarted by its unnatural buoyancy. With the clock ticking on our temporary importation documents and only a couple of days to spare, we knew we could afford little time to relax. We soon turned south to make our first attempt at crossing into Egypt.

Most of the dirt roads we'd taken the previous few weeks had been pretty demanding on our motorcycles' capabilities. After a particularly rough off-road ride through the Makhtesh Ramon crater, Sarah's bike developed a worrying electrical problem. Occasionally, at high speed, her engine would briefly cut out. When we arrived in the coastal city of Eilat, on the Red Sea, it became apparent we would have to diagnose and treat the problem before venturing into the Sinai desert. Shlomi's help proved invaluable when he called mechanics in Tel Aviv and translated the symptoms.

Thankfully, Sarah and I were riding almost identical bikes.

By interchanging parts, she narrowed down the cause to a faulty kickstand switch, probably damaged by a rock on the rough desert roads. Bypassing the switch solved the problem, but only after Sarah stripped her bike down to its bare essentials – twice – as Shlomi and I stood by and watched. We were impressed by how fast she completed this task on her second attempt while we offered redundant advice. She ignored our counsel and continued working through the heat of the day, dusty and sweaty and pleased with her efforts.

***

By now, we felt we were delaying the inevitable, and the time had come to say goodbye and enter Africa. On the day of our departure, we arose early and packed our bikes, moving slowly the way we did before any big decision. Shlomi accompanied us to the border, where we began the crossing procedures. As we handed over the necessary documents, we bade him a sad farewell – expecting never to return.

Little did we know.

Exiting Israel was relatively painless. After a cursory check of our paperwork and a few brief questions, the guards lifted the barriers and let us through. Two friendly Egyptian security officers asked a few questions about our intentions and waved us past a large sign listing all the prohibited articles we shouldn't be carrying. A mixture of fatigue and excitement prevented me from paying adequate attention to the sign, so when we arrived at the first checkpoint, the border guards had a field day going through my belongings.

Initially, the search was relaxed. When the call to prayer sounded, it seemed as though the office suddenly emptied. Only the few guards left behind carelessly sorted through our luggage. They appeared reluctant to touch any of Sarah's belongings, especially her clothing. But watching each piece of my gear go through the X-ray machine, something caught their eye.

They told me to empty every bag and explain the contents.

They became rather excited when they found my binoculars, forbidden in the Sinai. The discovery of a couple of knives in our tool kit had them summoning the entire hierarchy of security officers on duty. Finally, a couple of large plainclothes officers approached. In their mirrored aviator sunglasses, they casually enquired whether we had any intentions of getting up to mischief while in Egypt.

There were long moments of awkward silence when we wondered if this was the cue for offering *backsheesh,* an informal though widely practised form of bribery. Eventually, a senior official with more stripes, stars, and ribbons than we could count inspected us and our impressive arsenal and pointed us towards the customs outpost.

As we rode up to the front door of the customs building, a fierce-looking, older man in a wrinkled yellow shirt directed us to a parking space as he glared suspiciously at our motorcycles. We jumped off the bikes, removed our helmets, and gave him a warm smile, hoping to soften his sour mood, but he barked one word: "Carnet!"

Our spirits fell.

The *Carnet de Passages en Douane* can be a motorcycle traveller's biggest headache. After hearing stories from fellow overlanders who had managed without, we had deliberately decided not to travel with one. In essence, the Carnet is a document that allows travellers to temporarily import their vehicles into foreign countries without paying import duties or taxes. It ensures that the vehicle will be re-exported within a specified timeframe.

While the document was required to take a vehicle into a significant yet diminishing number of countries, it could be prohibitively expensive for travellers on a budget like ours. Egypt required an obscene deposit equal to many times the vehicle's value. A more affordable option was to purchase an insurance policy, which could cost thousands of pounds, especially when including Egypt. It was typically valid for only one year and often served to restrict travel rather than enable it. We had visited several online forums and heard of cheaper options being available at some border crossings, so we decided to attempt to go without one first and buy one only if necessary.

As the customs officer led me into his smoky office and sat behind his suspiciously empty desk, Sarah remained outside to watch over our bikes.

"Carnet!" he demanded as he reached his hand across the desk, expecting me to produce the document.

When I told him we were travelling without a Carnet, he rummaged through a pile of dusty folders on the floor, opening one and leafing through a random selection of paperwork before discarding the folder and choosing another. As he held up and examined the documents inside, each in a different language, he finally found what he was looking for, placing it on the desk where I could see it. Tapping it with his tobacco-stained finger, he repeated the word "Carnet" with increasing impatience.

"No Carnet," I replied. "What are the alternatives?"

"No Carnet, no Egypt!"

Again, an awkward silence followed as he glared at me, waiting for my response. Perhaps *backsheesh* would have helped my cause, but I didn't offer it.

He stood abruptly, piercing the air with the screech of his chair sliding backwards. As he dug in his pocket and produced a crumpled packet of cigarettes, he looked at me disdainfully before pointing towards the door.

He walked me briskly from the building, repeating his words, "No Carnet, no Egypt" angrily.

As I turned to leave, I asked him if I could purchase a Carnet locally, at which point he informed me I could buy one in Nuweiba, sixty kilometres to the south. When I offered to drive there to pick one up, he told me we would have to go back through Israel and on to Jordan before catching a slow ferry back to Egypt and into Nuweiba. And so it went.

I cursed him under my breath even though I knew he was just doing his job.

After a long, hot day of waiting, watching, and being watched, we ended up back where we started with little to show for our efforts. We returned to Israel, planning to do what we were told; that seemed to be our only option. If it failed, we would have to bypass Egypt altogether.

It was frustrating that alternatives existed to this archaic system, but some countries were slow or reluctant to adopt them. The money we would spend on the Carnet could have been better spent supporting local businesses rather than these fat cat insurance thieves. Many more overlanders might consider the African route were it not for Egypt's insistence on maintaining its grip on a derelict relic of the past, like a child clutching a favourite toy, even when it is battered and worn beyond recognition.

Denial of entry into Egypt, the gateway to Africa, put a significant roadblock in our way, but we hoped it would only be a temporary upset. Where we would go next remained a mystery, but if we knew the outcome of everything, it wouldn't be much of an adventure.

CHAPTER SIXTEEN

More Than I Could Bear

Personal Diary Entry: 5th December 2013

Day 206 (of 549)

33,177 km of 62,840 km (20,615 mi of 39,047 mi)

A strong wind whips the desert into a state of confusion. Atmosphere and earth combine, and a bright, brown halo hovers above a vague horizon, a hint of the blazing sun that lies beyond this sand-choked air. Obscured shadows blot the endless dunes; lonely trees, silent camels, dust-caked Bedouin driving parched goats across an unforgiving landscape. These are tough people scraping a meagre living from a land that reluctantly yields enough.

A storm is approaching. What will remain when the dust settles?

After our failed attempt to cross into Egypt, a feeling of despondency and weariness settled upon me.

It was a long ride back to Shlomi's house in the Negev region, and by the time we arrived, I felt terrible. Trying to ignore the aches and chills of an impending fever creeping into my bones, I barely registered Shlomi's delighted surprise to see us return. An unshakable weakness dampened my spirits and even my enthusiasm for a glass of wine when night fell and we reassessed our plans. I decided to turn in early. Sarah told me she was going to stay up longer and finish her drink with

Shlomi. Even through the fog of fatigue clouding my perception, I sensed a subtle hint of something odd.

I stumbled to the bedroom and collapsed onto the bed, feeling utterly drained. Sleep eluded me as I tossed and turned on the thin mattress, my monkey mind in a state of turmoil, leaping from one thought to another with little logical sequence. I must have slept eventually, as I awoke late the following morning, feeling only marginally better than I had the night before. Sarah was already up.

I hadn't felt that ill in a long time, and I was sure the stress of our latest upset was taking its toll on my immune system. At every border crossing we had attempted up to this point, there had always been some hope for success. This time, we had been refused with absolute certainty and finality.

It felt pointless to try again.

We had an established routine for dealing with the more complex border crossings where multiple offices were involved and the bikes would be unattended. I would gather our paperwork and deal with the logistics while Sarah would stay with the motorcycles and watch our gear. It was a system we had adopted and refined on our ride through South and Central America, and it appeared to work. I developed a sense for anticipating what documents would either help or hinder our progress and when to offer just the right amount of information required. One of the downsides of our system was that when it proved unsuccessful, responsibility fell solely on my shoulders.

Our denial of entry into Egypt presented a major roadblock to our progress. Alternative routes were few and far between. All eyes were on me; I had to figure it out on my own, and I didn't have any immediate answers. I knew a solution would present itself, but I didn't know when or from where.

I believed I was entirely at fault for the failure of our journey.

As I dragged my aching body out of bed, feeling defeated and lethargic, I resolved to find a bridge over this impasse. I shook my head to clear the clouds and prepared for a day of online searches and phone calls. I sat at the kitchen table, but everything felt oddly surreal that morning. The sunlight filtering through the window seemed murky and muted. I couldn't focus on the chatter between Sarah and Shlomi, and my ailing sense of taste turned breakfast into a joyless chore. A pounding headache made concentration near impossible, and I languished in a feeble state of stupor.

By midday, I had abandoned all attempts to solve the puzzle of how to enter Africa. Something was obviously wrong with me, and I needed rest. Sarah made me countless cups of tea, but her temperament was strangely different. I felt I had let her down and that she was disappointed with my inability to find a way forwards. As the day slowly faded into dusk and the temperature dropped, we ate a sombre meal that I tried to enjoy despite my condition. As I bade goodnight to Shlomi, Sarah, once again, said she was staying up to finish her drink.

I looked into her eyes and saw something I had never seen there before: deceit.

I shuffled off to bed, my mind in turmoil, questioning what I thought I had just witnessed. *Was my fever inducing some kind of paranoia? Was I imagining something that wasn't there?* I felt utterly miserable as I lay awake, my heart thumping and my bones throbbing. I waited for her to follow me to bed, and it struck me that I couldn't remember what time she had joined me the night before. After an eternity that couldn't have been more than thirty minutes, I rose and returned to the lounge.

The sound of me opening the door elicited a sudden response from Sarah and Shlomi; they jumped apart from being locked in a tight embrace on the battered couch. Their clothing was dishevelled; neither would look me in the eye as I sat down opposite them. Sarah muttered something about needing a shower and hurriedly left the room. I tried not to notice her open shirt.

Shlomi would not meet my gaze as I stared at him. We sat in silence as I waited for him to offer some explanation. I felt wave after wave of dreadful emotions wash over me. I didn't know what had happened between them, and I struggled to control my imagination.

Finally, I returned to the room I shared with Sarah. She lay awake with her back to me as I sat on the edge of the bed. "Well?" I asked.

No response.

I lied to her. I told her Shlomi had just told me everything. He hadn't. I told her it was now time she told me her side of the story.

And she did.

The revelation of her infidelity crushed me completely. It was late, and I was utterly exhausted, as my mind was spinning. I spent the rest of the night in a daze, drifting in and out of consciousness. As dawn finally broke, I silently packed my bike and rode off into the desert. I needed space and time to process what had happened, and the endless dunes provided plenty of that.

I lost myself in that sea of sand as a darkness consumed my thoughts.

I had devoted my life to this one person, and she had betrayed me. We had just faced one of the most challenging chapters of this journey, and I was suddenly full of doubt about the path ahead. All that we had worked towards looked like it was falling apart.

But I still felt our marriage was salvageable. The only thing I knew for certain was that I wanted to give her another chance. We're only human, after all; we make mistakes. We all have needs and desires. Sometimes, they lead us astray.

Like anyone else, I was full of flaws. Yet I was given grace again and again. Focusing on that thought, I contemplated how to respond to betrayal. I wanted to allow myself to be consumed with rage or at the very least, resentment, bitterness, or hostility. I decided on gentleness and understanding instead.

Choosing not to act upon our impulses requires discipline.

I had come of age in a nation scarred by animosity and vengeance, which I knew only led to a place of more pain and suffering. I had chosen a different way. *Could I stay true to that path and practice forgiveness for a wound that ran so deep?* Only time would tell, but I knew I had to try. When I returned to Shlomi's home, Sarah and I talked in our room and agreed to put what had happened behind us.

We left Israel, along with the mistakes of our past.

The country of Jordan welcomed us as we pressed onwards, its ancient landscapes a stark contrast to the turmoil brewing between us.

At our camp outside the city of Aqaba, we spent long nights discussing our future and our present. Yet, something was different. The person I thought I had known so intimately seemed to have drifted away. Sarah had changed; she became guarded and secretive.

I noticed she would lock the screen on her tablet whenever I approached, a small but telling action that gnawed at my trust. She seemed to have lost focus on the idea of going to Africa. Initially vague and unsettling, my suspicions grew sharper with each day as her behaviour fed my doubts. Clearly, she was struggling within herself, caught in a web of thoughts and emotions she chose not to share. When the weight of uncertainty became too much, I finally confronted her.

She confessed to still harbouring strong feelings for Shlomi, feelings she could neither suppress nor ignore, that were drawing her back to Israel. She had been communicating with him since we'd left, and they were conspiring to be together again. Even as we talked about forgiveness, she had planned to leave me.

In the heated argument that ensued, she unleashed a torrent of anger, accusing me of failing to meet needs she had never voiced. Her words cut deep, reopening old wounds while inflicting fresh ones, leaving me reeling under the weight of revelations I hadn't seen coming. The phrase she had once whispered so tenderly, "Always and forever," now felt like a bitter lie, its echo hollow and cruel, ripping through the fragile faith I had been struggling to rebuild.

It was more than I could bear.

CHAPTER SEVENTEEN

Buttered Side Up

Personal Diary Entry: 13th December 2013

Day 214 (of 549)

33,628 km of 62,840 km (20,895 mi of 39,047 mi)

Sarah has decided to follow her heart and be with her new lover, while my world lies in pieces. This morning she rode north, returning to Israel. Our journey together is over. The ease with which she could walk away from the life we built leaves me stunned and confused . . . alone.

After Sarah departed from Aqaba, I felt banished into a dark exile of solitude, alone with my emotions, paralyzed by feelings of loss. For three days, I waited, watching and listening. I wandered aimlessly through the city streets, hoping to see her face among the bustling crowds of strangers, desperate for one more glimpse of her radiant smile. Every time I heard the sound of a motorcycle, my heart raced and my spirits soared, but time after time, my hopes were crushed.

Minutes stretched into hours, each tick of the clock amplifying the longing within me. In a seedy hotel room, neglecting my hunger and fatigue, I stared endlessly at the walls, waiting for Sarah to reach out. As time passed, I slowly came to accept that our bond, our marriage, our connection had ended. The flame I carried within me that had once blazed with passion began to smoulder and quietly fade. On the morning of the fourth day, lost in my thoughts, I dropped a slice of toast on the floor. It landed buttered side up.

I smiled for the first time since this nightmare began.

I knew the moment had arrived when I had to let go and transform my anguish into something positive. It was time to summon the strength to face the unknown, embark on my journey of self-discovery, and find my path amid the ebb and flow of life's unpredictable tides.

The prospect of travelling alone brought with it a whole new set of considerations. It forced me to push myself out of my comfort zone, to open my eyes to what was around me, and actively seek out opportunities for social interaction. I was compelled to let go of the habit of turning to my partner for solutions or advice. I had no choice but to rely on myself more while asking for outside help at the same time.

I realised I needed to go home.

I had to be with people who knew me, who would accept me unconditionally. The suffocating weight of sadness that had settled upon me was not something I could handle on my own. With Christmas approaching and the rare opportunity to see my entire family together in Ireland, I made the selfish decision to squander a sizable chunk of my budget on a flight home. I hadn't been in Ireland for Christmas since the turn of the millennium thirteen years before, and the thought of being surrounded by people I love was too much to resist.

Through the Horizons Unlimited network of adventure riders willing to offer their help and support to fellow travellers, I was able to make contact with a new HU community member in Amman. The capital of Jordan and home to its largest international airport, Amman was roughly 325 kilometres north of Aqaba, where I had remained since Sarah's confession. After a few emails, my new contact offered me a place to store my bike near the airport while I returned to Ireland.

Methodically, I gathered my belongings and began to fasten everything back onto the motorcycle. Like the pieces of a jigsaw puzzle, every item had its place. Focusing on a practical task helped to distract me from my thoughts after so many days in limbo.

With the bike packed, I turned north, taking a detour to pass the last place I had seen Sarah. I wanted to extinguish the possibility that this had all been some terrible nightmare. Knowing I probably shouldn't, I rode towards the hotel where we had chosen to go our separate ways. Deep down, I still needed closure. I did not stop by the steps of the hotel. I only slowed down, glanced in that direction, felt my heartbeat stumble, and opened the throttle.

As I sped off into the desert, I did not look back.

Strong crosswinds pummelled my bike and pushed fine sand across the descending road as I dropped towards the Dead Sea. The fact that I was now riding through the lowest place on Earth was not lost on me. As the sun sank towards the horizon, the temperature began to plummet, and so did my thoughts.

It was an effort to keep my mind on the road, and I was thankful for the lack of traffic on my chosen route. During the long stretches when I was alone, I felt safe. But more than once, I felt seduced by the sight of an oncoming truck, realising I could instantly end my anguish by steering into the path of the approaching vehicle.

I had never before experienced such profound despair.

As my mind teetered on the edge of that black precipice, I reminded myself of the resilience that resides within the human spirit, the power to endure even the darkest storms. I clung to the belief that there was light beyond these heavy clouds. I was determined to keep moving, even though my heart ached unbearably. Each passing kilometre became a little victory, a declaration that I would not succumb to the allure of surrender. Though the pain persisted, I vowed to confront it head-on, to seek solace in the beauty of the world and the kindness of those I would encounter along the way.

My heart filled with a fragile hope.

To the west, across the Dead Sea, lights twinkled on the far-off shores of Israel. There, Sarah, my closest friend, my deepest connection, alive to everyone but me, was beginning her new life. Eventually, as the daylight faded and shadows swallowed all but my immediate surroundings, my entire world was reduced to only what my headlight illuminated. Darkness engulfed all but the tunnel of light ahead of me.

When a lonely police outpost appeared on the road ahead, I checked my speed and realised I'd been pushing my bike way too hard. I eased back on the throttle, but they'd already spotted me, and as I approached, they flagged me over towards the side of the road. I greeted them in my best Arabic, and they acknowledged my sorry effort with smiles and laughter. The three officers on duty gathered around my bike for the next thirty minutes as I attempted to describe my journey.

When I explained I was trying to get to Africa, they pointed over my shoulder and informed me I was going in the wrong direction. They were friendly, helpful, and curious about my thoughts on their country. Luckily, they showed no interest in issuing a ticket for my obvious speeding infraction. I continued on.

When fatigue finally overwhelmed me, I pulled off the main road and followed a dirt track down towards the water's edge. Amid the coarse rock and gravel, I found a flat space just big enough to roll out my sleeping pad.

***

The following dawn, I woke up stiff from the simple camp, though grateful for some sleep, and took in my surroundings in the soft light. The gently lapping water and birdsong lifted my spirits.

Soon, I was on my way. My route took me inland and up, climbing 1,200 metres before entering the capital, Amman. A recent storm had left the city with a heavy blanket of snow, and the locals were struggling to get all the roads open without any significant snow removal equipment. Most government buildings had closed their doors for the week, and those that were open provided limited services. My HU contact in Amman turned out to be a wonderfully kind Australian diplomat who met me as I entered the city and guided me to his part of town. Tellingly, they had an armed guard posted at their main entrance.

Founded in 7000 BC, the city's streets seemed to follow no logical pattern, and I was glad for the help in finding my way around. I spent the evening enjoying the kind hospitality of my new friend and his partner as they opened their home to me, a complete stranger, and made me feel welcome and cared for.

After so many days alone, I was yearning for company.

That night, we dined at the nearby English pub and met up with some local adventure riders keen to share stories and advice from their travels at home and abroad. I was able to escape my dark thoughts for an evening.

The following day, I locked up my bike and headed to the airport by taxi, stopping briefly at the Royal Automobile Museum. An armed soldier at the entrance informed me that the museum was closed because of snow but, after I stubbornly refused to leave, the curator emerged and we talked. When I told him of my journey, he opened the doors, and I had the entire building to myself.

Persistence pays off.

Inside was a collection of vehicles gifted to the king of Jordan throughout the country's short history, including over fifty motorcycles. As the expansive roof groaned and creaked under the weight of the snow, steady water drips splashed onto the exhibits and the surrounding floor. Such a heavy dump of snow was an unusual event in this part of the world. I was startled that the museum had not considered how to protect its collection from the elements.

I wandered around, appreciating what there was to look at. As I readied to leave, I stopped to gaze again at what had initially caught my eye. Parked at the entrance was a working replica of the world's first-ever motorcycle, the Daimler *Reitwagen*, assembled in 1885. I wished I had brought my bike along to compare the two.

How things had changed in such a short space of time.

A day later, as I stepped off the plane in Belfast, an icy wind quickly found its way into every gap in my inadequate clothing, but the loving smile and strong embrace of my father brought the comfort I had longed for. To be home with family and friends during such a difficult period was just what I needed. I still felt broken and raw, but with time, I knew I would heal; the wounds would close and the memories would fade.

Turning my back on a decade of devotion to one person was the hardest thing I could imagine, but I had to do it. What other choice did I have?

CHAPTER EIGHTEEN

Episode of Déjà Vu

Personal Diary Entry: 8th January 2014

Day 240 (of 549)

34,095 km of 62,840 km (21,310 mi of 39,047 mi)

They say it only rains twice a week in Ireland, from Monday to Friday and then from Saturday to Sunday – only twice. But not even the cold, wet weather could dampen the spirits of my family and friends as we gathered in Strabane to see out the old year and welcome in the new.

Coming home for Christmas was one of the best decisions I have ever made. I arrived feeling discarded, like an empty shell, devoid of hope and lost in a sea of confusion. Yet, within hours of landing, I began to feel the calm stability I so desperately needed.

The future I had expected was no more, and the narrative of my past was up for review as Sarah, very publicly, cruelly rewrote it, telling her side of the story. It was almost convincing enough for me to question my memories. Utter fatigue plays foul tricks on the mind, and had it not been for the support of my family and friends from around the world, I would have begun to doubt my sanity. Bathed in a deep, warm pool of love and understanding, I finally began to find peace.

I was home.

Ireland is a magical place. Of course, I am biassed, but the compassion and hospitality of its people are second to none. Grounded in reality by the shared experience of a troubled past, we did our best to live normal lives under abnormal conditions. The quick wit and the dry humour are merciless, but beneath the tough appearances, there lies a kindness and warmth that would melt the coldest of hearts.

I've always suspected the Irish climate has played a pivotal role in forming our national psyche. Where better to avoid the inclement weather than drinking tea and telling stories at the kitchen hearth or sharing music and poetry around the fireplace of the local pub? I could think of no other place I would rather be during those dark days than with my family, surrounded by the familiar sights and sounds of my youth.

Gathered under one roof, the entire Anderson clan set the scene for several days of eating, drinking, chatting, singing, joke-telling, and general merriment, followed by more of the same. I can't remember how many cups of tea I drank during the days I was home, but that may have had something to do with the amount of fine whiskey I consumed in the evenings. Our time together constantly reminded me of just how fortunate I was to have such a close, supportive family by my side when I needed them most.

As we joyously rang in the new year together, I pondered on my one thought most present: with each new year comes a new beginning.

***

I'd already pushed my flight back once to spend more time with family, but I knew I would have to get back on the road eventually. So, I forced myself to pack up and with a heavy heart, I got on the plane. Returning to Jordan, home of such painful memories, was challenging; however, my spirits rose as I entered the airport's arrivals area in Amman.

A smartly dressed driver held a sign that read "Mr. Irish Dave."

The overwhelming hospitality of my wonderful new Australian HU friends greatly eased my return to the road. They had lovingly taken care of my gear and motorcycle while I was in Ireland.

As exhausting as the journey had been, I spent a sleepless night toiling over what my next move should be. *Should I throw in the towel or keep moving?* Since the breakup with my wife, several attractive opportunities had presented themselves, but deep down, I knew it would be cathartic to finish what I'd started, even though it had cost me so much. I made the decision to explore several options, the first of which involved attempting to transit through Saudi Arabia to Yemen, where I could, hopefully, catch a boat to Djibouti.

Obtaining a visa for Saudi Arabia was said to be notoriously difficult – unless you were an arms dealer – and my first visit to their embassy proved to be a fruitless waste of time. The staff seemed reluctant to deal directly with a Westerner, so I was refused entry into the building and directed towards a visa agency "on Airport Way, near bridge two." Directions could be complicated in Jordan where people rarely used full addresses. It felt as though they did not want to disappoint me with a flat "no," but they would prefer it if I went elsewhere.

Bureaucracy and fatigue do not go well together.

I meekly agreed to move along and attempt what I thought would be the simpler task of extending my permission to keep the motorcycle in Jordan for a few more days, during which I'd work on securing a Saudi transit visa. On entering Jordan, a temporary importation licence had been granted for my "foreign vehicle," allowing me to stay for up to thirty days. It was about to expire, so I took a drive to the border to renew or extend my time limit.

Approaching the border, I pulled over at the first of three checkpoints on the Jordanian side and began explaining my predicament. The police officers were friendly enough, but they shook their heads together after inspecting my paperwork and repeated a phrase I had heard many times before on this trip.

"You have problem."

Phone calls began back and forth between the various border agencies, and as the sun set, I wondered if I had made a mistake by playing by the rules. Finally, I was allowed to proceed to the next checkpoint, where the entire procedure repeated itself in a surreal episode of déjà vu.

It was dark when a senior-looking police officer approached and informed me that I would have to leave the country and return the following day to extend my stay. I'd left most of my gear in Amman, expecting a simple case of paying a small fee, but the customs officials were determined to do it all by the book.

When I refused to leave, the senior police officer suggested I overstay my welcome, remain in the country illegally, and pay a small fine upon leaving – considerably less than the cost of the extended import licence.

Turns out, breaking the law was cheaper.

***

The following morning, I set out to track down the mysterious Saudi visa agency with the vague directions given to me. Assuming that the address on Airport Way may be related to the airport's location, I traced a route along all the possible roads that led out of the city in that direction, but to no avail.

I returned to the Saudi embassy and asked for a more accurate location. I spoke to a new set of officials who gave me a completely different set of directions, but by then, the day was wearing on, so I decided to give it another go the next day. Returning to the Amman apartment, with its armed guard stationed outside, I spent the remainder of the afternoon working with my host on his bike, repairing some damage to its front end after a mishap in the desert.

After another sleepless night, I pursued the Saudi visa option again for the third day, finally tracking down the agency that processed applications for foreign visitors.

In Jordan, they often describe how well-connected a person is by the term *wasta*. The king of Jordan would, presumably, have the most wasta, and I, as a foreigner, would have very little *wasta*. It was all about who you knew in positions of power, and achieving the simplest of tasks without adequate *wasta* could be next to impossible.

The staff at the visa office were most helpful, and initially, quite positive about my chances of obtaining the seventy-two-hour transit visa required to race across their enormous country and into Yemen.

That was until they asked what type of vehicle I would be driving.

I had assumed that would have been obvious – after all, I was wearing a full riding outfit and carrying a crash helmet. Yet, as soon as I mentioned the motorcycle, attitudes changed and my lack of *wasta* sealed the deal.

Saudi Arabia shared an odd paranoia with much of the Middle East when it came to motorcycles, and they would not issue a transit visa for all motor vehicles of the two-wheeled variety. I was now dealing with yet another rejection, and my options were gradually falling by the wayside. I knew there was a solution that would present itself in time if I was patient.

With each rejection comes time for reflection.

As I dejectedly rode away, I thought of another time, long ago, before my heart was broken by the woman I loved, back when we had just married. It was a moment when Sarah and I encountered a major setback, and everything had magically worked out.

On our two-and-a-half-month honeymoon one-motorcycle journey from Colorado to Chile, we encountered our fair share of setbacks. One memorable roadblock occurred after the punishing ride through Mexico. To save money, we chose the scenic, rural routes to avoid expensive toll roads. It initially seemed like a charming idea, passing through quaint towns and villages, each with its distinct character. But these municipalities had a peculiar way of greeting travellers – with aggressive speed bumps – and not just one or two, but entire battalions of them, placed strategically to ensure you never forgot you were entering or exiting a village. The pattern repeated as predictably as the sunrise, each bump adding a little more wear and tear to us and the bike.

Entering Guatemala, with its smoother roads, finally allowed me to open the throttle so we could enjoy the full potential of our powerful machine. Right away, I sensed something about the bike that felt different. A strange vibration I hadn't noticed before was so subtle that I assumed it was just my imagination playing tricks on me. As the kilometres rolled past beneath us, it became harder to deny that something was wrong. I stopped several times and examined the bike, checking the tyres for damage or insufficient pressure, hoping it would be a simple fix.

Deep within me, a suspicion stirred, hinting that the problem was more serious.

Finally, on the outskirts of the city of Quetzaltenango, I pulled over and conducted a thorough inspection. On the side of the road, with our luggage spread around us, I lay in the dirt and found what I hoped I wouldn't. A tell-tale trickle of thin oil oozed from the rear hub of the bike. I had read about this symptom, and my heart sank as I knew what it meant. The final drive bearing was failing – a critical component that supports the rear wheel as the drive shaft transfers its power from the engine to the axle, and its replacement was considered so complex that BMW recommended using one of their expert technicians for the job.

"Can't you fix it?" Sarah asked.

"I don't think so," I replied. "Not with the tools we are carrying. This is going to be a tricky one."

I had researched the location of BMW motorcycle workshops before leaving the US; the nearest one was in Mexico City. I would have to dismantle the rear end of the bike and take it back to Mexico via public transport to have it repaired. I knew the cost would take a significant chunk of our limited funds. Plus, it would involve leaving Sarah in an unfamiliar city with the rest of the parked bike while I was gone. My mind ran through all the alternative solutions, but none of them seemed viable. We were in a new country, our Spanish was terrible, we had a deadline for this trip, and we had stretched our budget thin as it was. I consulted the mechanic's manual I had thoughtfully packed, but it only confirmed that this job was best left to the experts. I was out of ideas.

As I ran my oily fingers through my messy, curly hair in frustration, a small pickup truck pulled over across the street from us.

"You have problem?" the driver shouted above the noise of the traffic.

"Tengo problema grande!" I yelled back in my best Spanish.

"No problem, follow me." He motioned with his hand and flashed us a quick smile.

Sarah and I exchanged wary glances. "What do you think?" I asked her.

"I don't know . . . this feels kind of weird," she answered.

"I'm all out of ideas right now, and I'm willing to take a chance if you are. What's the worst that can happen?" I offered. "On second thought, let's not think about that."

I turned back to the driver and cupped my hands around my mouth to amplify my voice. *"Un momento, por favor."*

He gave us a thumbs-up gesture, and we quickly packed our gear back onto the bike. Once we were mounted and ready to go, I returned his thumbs-up, and we pulled in behind his vehicle as he entered the flow of traffic. He suddenly turned off the main road into a quiet side street, then another and another. We followed him through a maze of narrow alleys.

Just as I began to question the wisdom of our decision, we stopped outside a nondescript compound protected by a heavy steel gate and concertina wire atop its tall brick walls. Somehow, we were less than 500 metres from where we had started this mysterious pursuit. Our new acquaintance jumped from his truck, approached the gate, and banged it several times, tapping out a distinct pattern. A vicious-sounding dog responded from within. Sarah tightened her grip around my waist. A heavy latch was unhooked and the door slid open. A short, barrel-chested man with thick glasses greeted us with the warmest smile, and as I felt Sarah's grip loosen ever so slightly, I could breathe again.

Within his tidy compound stood a row of immaculate BMW motorcycles from different years, each in various states of repair. We rode our bike inside. After we parked and dismounted, the pickup driver introduced us to Roberto Ascoli, president of the BMW Motorcycle Owners Club of Central America. With one look at our bike, he saw the trickle of oil beneath the rear hub.

"Ha!" he exclaimed. "I know this problem."

He excitedly ran off into his workshop, where I could hear him rummaging through drawers and boxes. Soon, he returned with a bearing raised above his head triumphantly. "You need this."

I gave Sarah an incredulous look, and we giggled at the absurdity of it all. I wondered how any of this was even possible.

It didn't take Roberto long to establish that our Spanish was appalling, and he easily talked us into staying in his city to take lessons. We had already been planning to do this somewhere along the way, and when he suggested doing so right there in Quetzaltenango, we consulted our guidebook for the best places to study Spanish; this city was top of the list. It had a welcoming attitude towards foreigners and offered an affordable cost of living and an immersive cultural environment.

Plus, this peaceful city in the fertile green valley of the Western Highlands of Guatemala had so much more for two honeymooners to enjoy: volcanoes to hike, hot springs to soak in, and beautiful neoclassical architecture to admire. We thanked the pickup driver for his divine timing and agreed to leave our bike with Roberto while we immersed ourselves in a week of intensive Spanish lessons.

It was another reminder that sometimes the road leads you exactly where you need to be.

As I thought about that honeymoon trip at the beginning of my marriage, which was now at its end, I realised something. Just like the breakdown of our bike turned out okay, I'd turn out okay, too.

CHAPTER NINETEEN

Faintest Flicker of Optimism

Personal Diary Entry: 25th January 2014

Day 257 (of 549)

37,203 km of 62,840 km (23,117 mi of 39,047 mi)

It feels as though I've suddenly woken up to find myself middle-aged and single. For the last 10 years, I felt lucky to have found a kindred spirit, someone to share my hopes and dreams with. I don't know at what point she turned off her love for me, but I realise what has gone can never be recovered. I'd rather she be happy with someone else than unhappy with me. I'm not sure if I will ever love or trust again – it's still too soon to say. To consider it, even now, feels like a betrayal. I'm sure that will change with time.

Right now, I need to recalibrate who I am and reconsider what direction I want to take my life. I want to learn to put myself first again. It's time to begin the process of moving on.

Chasing paperwork down for two weeks in Jordan had involved some frustrating dead ends. At one point, it felt as though I would never be able to leave the country, stuck in the place where my life fell apart.

I decided to make the most of it.

While getting one rejection after another, I went on some incredible excursions, riding alongside the local adventure bikers of Jordan.

Numbering barely a dozen, it was a close-knit group who enjoyed the varied terrain of the vast landscape.

From open desert to scenic twisting asphalt to formidable dirt tracks, they had it all here in Jordan, and they weren't afraid to stretch their bikes and bodies to the limits of their capabilities. I put many more kilometres on my bike than I had expected to, and my off-road abilities improved dramatically thanks to some helpful tips from my fellow riders.

It was truly a biker's paradise.

Endless valleys bisected the country's mountainous central region, where the Bedouin people still roamed freely. Dotted with ancient ruins dating from over 7,000 years ago, it represented vast swathes of human history, indicating the length of habitation. Highlights included finishing a long day of on- and off-road riding with a relaxing soak in some 2,000-year-old Roman hot pools and breaking fresh trails through the eastern desert to ancient fortifications that stood lonely and abandoned throughout the region.

Then, finally, I had a breakthrough and obtained the coveted Carnet de Passages for the motorcycle, which would allow me to enter Egypt. After almost a month in limbo, with multiple failed attempts, I was finally going to set foot onto Egyptian soil – or sand.

With a mixture of sadness and relief, I packed my bike with all my worldly possessions and, for the final time, rode the Dead Sea road to the port city of Aqaba. I was going to miss Jordan and the genuine kind-heartedness of its people. Without their help, my journey may have come to a stuttering halt.

As the road wound its way down from the Amman plateau to the Dead Sea basin, a subtle haze of the finest sand filled the air. The jagged profiles of the central mountains that flank the valley appeared as a succession of fading ridge lines softly vanishing off into the distance as far as the eye could see. I stopped at the side of the road to truly absorb the scene and that moment – starting a completely different type of trip than the one I had set out on with Sarah.

I realised I didn't disappoint Sarah when I couldn't get us into Egypt. She didn't want to go to Egypt at all. She wanted to stay back, create a new life with someone else, and have me go on without her. Well, I was about to do just that. I was about to finally enter the elusive country of Egypt – alone.

As I stood at the crossroads of uncertainty, the future stretched out before me like an uncharted territory. While I had only an obscure view of the path ahead, I felt the faintest flicker of optimism permeating the air. From the depths of the turmoil and heartache, I knew things could only improve.

I was eager to put the recent past behind me, so I raced off at full throttle.

***

Equipped with what I hoped was the required paperwork for legal entry into Egypt, I boarded the ferry from Aqaba to Nuweiba.

I secured my bike to the cargo deck and settled in for a long, sleepless night as the boat slowly chugged its way across the Red Sea. Why this overcrowded, stinking hulk of a vessel couldn't make the crossing during daylight hours remained a mystery. In the chilly pre-dawn haze, it dumped its tired cargo on Egyptian soil long before the border officials awoke and prepared for the sudden influx of traffic.

Knowing what to expect after my previous failed attempt certainly helped to ease the frustration, and I quickly commandeered the services of a local police officer with the promise of a little *backsheesh* for his help. Piece by piece, we put the jigsaw puzzle of paperwork together – acquiring an importation permit, obtaining vehicle insurance, getting Egyptian registration plates fitted to the bike, applying for a driver's licence in Arabic – all while attempting to satisfy the security detail on duty that I did not pose a threat to their country. Everything would probably have been relatively straightforward had it not been for the abandonment of ongoing renovations to the port facilities since the recent changes within the country's leadership. It was a vast and busy complex in a state of confusing turmoil.

At one point, I had to bang on the door of the vehicle inspection officer, who would ascertain whether or not my bike contained any hazardous goods. After five minutes of knocking, a tired groan emerged from within, and another five minutes passed before a sleepy-looking officer in a nightshirt cracked the door and peered at me, then the bike, and then me again before reaching out his hand for the required fee.

The door slammed shut, and another ten minutes went by, as he reorganised his furniture looking for the correct form. Finally, the door rattled, and through a thin crack, my certificate emerged, confirming that, indeed, I posed no threat. Eventually, after several tedious hours of more nonsense, an officer directed me towards the exit, which I assumed was my cue to leave.

An incredulous grin spread across my face as I rode my motorcycle between the heavily armed soldiers standing guard at the final gate. At the very last minute, I had to slam on the brakes as one more official stepped into my path, and we, again, went through the reams of paperwork that I had stuffed into the numerous pockets of my motorcycle jacket.

Recent sinister events in Cairo and the Sinai had, understandably, left the army and police in a state of heightened alert. After a long, slow inspection and much staring and shrugging of shoulders, he stepped aside. "Welcome to Egypt," he said as he gestured for me to go.

I wouldn't technically enter Africa until I crossed the Suez Canal, but I felt like I had passed another significant milestone. I would no longer have to keep banging my head against the brick wall of Egyptian bureaucracy. Although there were still almost 10,000 kilometres between Cairo and Cape Town and many more borders to cross, it seemed the most complicated one was behind me. One by one, countries were slowly moving away from the Carnet de Passages system, and soon, that archaic piece of paper would become a relic of the past, where it belonged.

I spent the first few nights in the Sinai at the crumbling but quaint beachside resort of Dahab, on the eastern side of the peninsula. Accommodation was cheap, so I opted for a deserted hotel overlooking the ocean. I was even allowed to bring my bike inside the compound and park it right outside my room.

As the only guest, the staff treated me like a VIP.

Immediately befriending me, the Bedouin staff was soon parading me around the town, introducing me to more people than I could possibly remember. They made me feel like an honoured guest for simply making the effort to stop by.

Everyone seemed to be suffering because of the lack of tourism, and many businesses had closed since the revolution three years before. The shops and restaurants that had survived were mostly empty and desperate for any trade, but the instability in Cairo had scared off the majority of tourists. What would become of Dahab remained to be seen.

Less than fifty metres from my hotel room was a golden beach where the ocean floor dropped off over a precipitous shelf, the edge of which was home to a spectacular coral reef teeming with life. I rented some snorkelling gear and a wetsuit and spent a long time swimming along the edge of the reef, gazing in awe at the variety of fish and coral so close to the shoreline.

A sea turtle swam next to me for a few minutes, appearing to enjoy my company as we examined the coral. The vividly coloured fish of all shapes and sizes seemed oblivious to our presence as they scoured the reef for tiny morsels of food.

***

Through my friends in Jordan, I had made contact with an Egyptian biker, Nabil, who lived on the western side of the Sinai. He agreed to meet me in Dahab, and together, we rode to some of the region's hidden gems.

The Wadi Gnai provided an opportunity for some strenuous off-road riding in loose sand, but as always, anything worthwhile takes a bit of effort. The serenity of the canyon was only disturbed by the quiet ticking of our cooling engines as we stopped to explore the dry creek beds that fanned out from the valley floor.

We rode north to the infamous Blue Hole, a diving mecca that attracted explorers from around the world because of its exquisite location and the wealth of sea life inhabiting its pristine coral reef. Plaques adorned the walls on the walk to the lagoon, honouring the divers who allowed the reef to draw them deeper than they should have gone. As we snorkelled along the surface, I could understand the allure of wanting to see more down below.

We left Dahab early one morning and rode to Saint Catherine's Monastery, an Eastern Orthodox Christian site from the seventh century still used today. Being the oldest continuously inhabited Christian monastery in the world, it claims to be the exact location where Moses encountered the burning bush.[14] We stopped briefly to investigate the complex before making our way towards the inland route to the Western Sinai.

Security checkpoints dotted the roads in this region.

Barricades crossed the road every forty or fifty kilometres, and heavily armed police and soldiers patrolled the temporary installations. As we crossed the final checkpoint before entering the heart of the Sinai, a guard told me I could not proceed. Being a foreigner, I was considered a high-value target for the Bedouin, who would occasionally kidnap tourists.

A tit-for-tat game of arrests and kidnappings had recently evolved in the area. When the police arrested one of the Bedouin, the Bedouin took a tourist as a hostage to negotiate an exchange. A couple of tourists who had such an experience admitted afterwards it was a highlight of their trip to Egypt. The Bedouin treated them like honoured guests and were polite and gentle, then made the swap, releasing both parties. The police were unwilling to risk a repeat occurrence, so they would not let me proceed, forcing Nabil and me to take the much longer coastal road around the Sinai.

It was a spectacular route with a strong headwind that made for an exhilarating and exhausting ride.

Night fell as we pushed on towards Nabil's home in Ras Sedr, and I had my first experience of driving in the dark in Egypt. Many bikers had warned me of the unusual practice where locals would drive at night without headlights. My motorcycle's headlamp automatically stays on constantly, and during the day, oncoming cars would often flash their lights to remind me that I had forgotten to turn it off. At night in Egypt, cars would creep up unseen behind a vehicle until a quick flicker of light indicated their intention to pass. While not the safest practice, it served as a good reminder to be off the roads before nightfall.

It was around midnight when we finally arrived in Ras Sedr, a peaceful resort and kite-boarding mecca on the shores of the Red Sea just south of the entrance to the Suez Canal. Heavily laden cargo ships slipped by quietly on the horizon, a steady procession of supplies going to and coming from the Mediterranean. While the revenue from the Suez Canal supported the government and the military, many working-class people relied on tourism as their primary source of income. However, the way the international press portrayed Egypt dramatically impacted visitor numbers.

The news outlets reported only on the sporadic violence that occasionally rocked the country, such as the protests leading up to the third anniversary of the revolution that had left several people dead. As a result, tourism numbers were well below average, and the economy was suffering. The advice I was receiving from locals was to avoid the capital, my next destination, for a little longer until things had settled down.

I decided not to let this warning stop me.

The time had come to bid Nabil farewell and continue alone. This part of the journey had come with plenty of challenges. I was happy to have had a new friend along for part of the ride, but I missed being able to share my experiences with Sarah. I longed to chat with her through the intercom on the quieter sections of the road the way we used to.

The inside of a quiet crash helmet could be a lonely place.

As much as I tried to prevent it, recollections of the last few months would begin to appear, and I slipped into a deep melancholy over what

I had lost. I thought of how the absence of Sarah's love for me felt like something essential for my survival suddenly torn away from me. I remembered how, when we parted ways, my love for her had no place to go. And afterwards, how dark my thoughts became to the point where I didn't think it was a good idea for me to be alone.

I struggled with the "what ifs" and "if onlys" that would surface during those moments. After I allowed my mind to go there, I'd desperately cling to the memories of the past, all while knowing how detrimental it was for me to do so. I wanted to be present, not always stuck in my head. I didn't want my bitterness to leave a bad taste in my mouth and spoil the rest of this trip.

I didn't know how to heal, but I was trying.

CHAPTER TWENTY

Away from the Chaos

Personal Diary Entry: 8th February 2014
Day 271 (of 549)
39,397 km of 62,840 km (24,480 mi of 39,047 mi)
I will move on, I will grow, I will learn. I will spend time with myself and try to discover who I am. As impossible as it seems now, I will recover and rebuild, I will forgive. I am nervous about what lies ahead, where the road will take me, and whom I will meet along the way. With no schedule or agenda, I can follow my own heart.

I feel it is time for the real journey to begin.

Travelling alone made me more approachable and the interactions with locals were much more enlightening and meaningful. People seemed more willing to initiate conversations and offer advice and hospitality.

It speaks volumes about human nature that when we are alone, people feel the need to offer company. Some wanted to enquire about my journey, others wanted to practise their English, and with all, a smile was never far from the surface. Plus, something miraculous happened during my travels: the stranger and I always discovered we had more in common than not.

As I turned my bike towards Cairo, I decided not to let my sadness make me unpleasant when crossing paths with these kind and curious strangers. I vowed to do my best to remain open, trusting, and eager to connect with them the way they were with me. Likewise, I hoped to learn to approach life the same way.

I wanted to surrender myself to whatever lay ahead of me.

Cairo, Egypt's dusty, bustling capital and Africa's largest and fastest-growing city, somehow seemed to function under the strain of warring political factions, limited resources, and a faltering economy. Awestruck, I balanced my bike next to the curb and gazed at the thriving metropolis built upon a historical legacy stretching back millennia.

The resilience and resolve of its enormous population inspired me. I sensed the energy around me as almost palpable, my entire body tingling with anticipation. The streets were impossibly crowded, and the notorious traffic kept moving as drivers squeezed through the clogged arteries from the city's thumping heart to its ever-expanding periphery.

I was apprehensive about plunging into the chaos.

Yet, when I released my toe-hold on the curb and mentally shoved myself into the commotion, I learned to accept Cairo's rhythm and go with the flow. Soon, it became masochistically pleasurable to ride through the narrow gaps that briefly appeared between overloaded trucks and smoke-belching buses. The screeching of brakes and the blaring of horns were ubiquitous throughout, adding to the atmosphere of urgency and mayhem.

How anyone could live amid this urban extravaganza amazed me, but they did and in their millions. I had heard all kinds of estimates as to what the true numbers were, but many seemed to agree the population of Cairo and its suburbs was close to thirty million.

What made everything seem even more hectic was the current state of heightened police and military activity, where many streets were blocked by tanks and barbed-wire barricades. I was astonished that what should have caused complete gridlock didn't deter the locals from going on with their daily commutes and activities. But I did sense a general disheartenment around me. After so many years of resistance, the residents were reminded daily that the military were still in control.

Many feared they always would be.

The political situation here was complex. The revolution that had ended the authoritarian reign of Hosni Mubarak three years earlier appeared to be approaching a full circle. A return to the way it had been prior to the uprising now seemed inevitable. The optimism that had gripped the country until recently had mostly evaporated, and the brief experiment with democracy seemed to be tragically backfiring. The corruption that fuelled the protesters' anger was so deeply ingrained that the entire system would need to be torn down if genuine change were ever to take place.

Speaking to people who took part in the events in Tahrir Square, it became clear that the ideals they fought so bravely for had come to nothing. A revolving door existed between the military and most branches of government, with many retired officers going on to assume key roles within the country's fragile infrastructure. When the Muslim Brotherhood took control, rolling power cuts plagued the country, as well as water and fuel shortages – much of it engineered to weaken the newly elected leadership.

A population could only tolerate disruption like that for so long. Soon, the strain manifested in further protests until the military staged a coup and regained control, having never really conceded it. They now controlled the media with such an iron fist that any dissent was dealt with harshly. Arrests were common and newspapers and television stations had been shut down when they challenged the status quo.

Bassem Youssef, the wildly popular satirical news reader, had been taken off the air for openly criticising the government. Facebook and Twitter had been targeted with censorship, and the military retained the power to cut off all forms of external communication. Much had been learned from the methods employed by the original protesters, and the new regime was taking all precautions to prevent a repeat of the events of three years before.

At this point, most of the people I had spoken to were tired of the situation and willing to accept a stable but corrupt system over anything proposed by the Muslim Brotherhood. They just wanted things to go back to the way they were before.

The people wanted peace.

Yet, at no point in my travels through Egypt did I feel threatened or uncomfortable. While there were plenty of curious stares, when I responded with a smile, I often received a smile in return. From the *felucca* captains to the taxi drivers, they certainly hustled hard for my attention when I was on foot, but two or three polite refusals and they would invariably retreat to look for business elsewhere.

Arriving in Cairo after the long ride around the Sinai, I was promptly greeted by another Horizons Unlimited member, a keen overland enthusiast. A mild-mannered, good-natured teacher by day and an intrepid explorer during his time off, Sam could often be found roaming the infamous Sand Sea in his trusty Land Rover, "Stella."

We spent our first night exchanging stories over curry and cold beer before returning to his ground-floor apartment in a grand old colonial house in the suburbs of Maadi. His last trip into the Western Desert, along some of the routes used by the Long Range Desert Group during World War II, had involved getting shot at by the Egyptian army, stumbling across an active smuggler's cache, and rolling one of the Land Rovers – exciting stuff!

Using Maadi as a base, I dropped into the local BMW motorcycle dealership for a chat about suitable rides while in Egypt. The staff there invited me to stop by the Egypt Motorcycle School (EMS) afterwards to meet some local riders and take care of a few routine maintenance procedures on my bike.

When I dropped by the school later in the week, I was extremely impressed by their operation. They ran a fleet of smaller Chinese bikes to introduce beginners to basic skills and a couple of bigger Japanese road bikes for the more advanced students. All lessons came with helmets and body armour included, two things I rarely saw on the roads in Egypt. Nothing else like this existed in the country, and it was encouraging to see the level of professionalism with which they approached educating young riders.

While there, they invited me to give an informal presentation about my journey.

I couldn't refuse.

Afterwards, when the team at EMS learned that my bike was overdue for an oil change and some maintenance on the chain, they insisted on supplying everything I needed while completing all the work on the bike. When all was done, they put out a spread of food that could have fed a small army, and before long, we had made plans for some desert rides while I was in the area.

During my stay in Cairo, I couldn't resist a visit to the Pyramids of Giza. I made my way out to the edge of the city, where the desert plateau met the suburbs, and spent a day wandering among these amazing structures. Nine pyramids occupy the Giza site, along with numerous other lesser temples, but the three pyramids of Khufu, Khafre, and Menkaure and the iconic Sphinx made the site so well-known.

The last remaining structures of the seven ancient wonders of the world served as mausoleums for the pharaohs on their path to the afterlife. Closer inspection finally satisfied a curiosity that had puzzled me ever since I was a young child. *Was there such a thing as the Sphinx's sphincter?*

Standing at the base of each structure, I found it hard to imagine how they had ever built the pyramids. I pictured the physical effort required to put one block in its place – never mind the thousands that went into each pyramid – and all without modern machinery! Somehow, each one was still standing strong after over four millennia.

They sure don't make them like they used to.

Imagining all that physical effort made me quite thirsty, so I retired to the nearby Mena House Hotel for a cold beer. It was a rather fancy establishment and way out of my league, but with a spectacular view of the pyramids, I couldn't pass it by.

The following day, I ventured into the city centre via the Cairo Metro. One of only two fully-fledged metro systems in Africa, it was said to move around a billion passengers annually. I was impressed when I could hop on for the minimal fee of one Egyptian pound – the equivalent of eleven to twelve British pence.

Once in the city, I sought out the Egyptian Museum for its extensive collection of pharaonic artefacts, including the evocative death mask of Tutankhamun. They confiscated cameras at entry, so the postcard vendor outside saw plenty of trade.

After the museum, I attempted to stroll through Tahrir Square, but the military had the entire area cordoned off with barbed wire and tanks. All signs of the recent protests had been erased, and much of the graffiti that once adorned the surrounding buildings had been hastily painted over. The absence of civilians felt ominous, as did the presence of so many heavily armed soldiers. Locals appeared to be making a conscious effort to avoid looking in the direction of the square for fear of attracting the attention of the brutal state security service.

From Tahrir, I wandered through the heart of the city towards the grand bazaar of Khan al-Khalili, a warren of alleyways selling a prodigious variety of locally made goods. The smells that wafted through the dark, narrow passages combined the sweet fragrance of bubbling sheeshas with the tangy scent of fresh spices. The market was thriving with activity, and there was a noticeable absence of the tourist tack I'd found in so many other city bazaars.

This one felt like the real thing.

The cornucopia of sights, sounds, and smells triggered a deep desire to eat. My stomach grumbled, reminding me I hadn't consumed anything since early morning. It was now mid-afternoon. I suddenly felt weak and a little overwhelmed by all the tantalising choices on display. I let my nose decide and was drawn towards a stiflingly hot grill selling sizzling kofta kebabs composed of some kind of mystery meat coated in a combination of unfamiliar spices. I hungrily devoured one and immediately asked for another before noticing another vendor selling freshly fried falafel. I was suddenly even more ravenous and was convinced I could spend the remainder of the day sampling every delicacy the market had to offer.

After satiating my hunger, I felt much better. I allowed the intoxicating atmosphere of the moment to wash over me and cast myself adrift in the steady current of locals flowing through the vast market.

The sound of voices filled the air, gossiping, laughing, haggling, all in a language I couldn't comprehend. Coins jingled back and forth as deals were made and goods exchanged. Traders waved at me to come and inspect their wares and I respectfully declined with the few Arabic phrases I'd learned. I didn't mind purchasing consumable goods but I did not need any souvenirs. It was joyful to lose myself amid the bustling crowds, and brushing shoulders with complete strangers reminded me how much I was missing the human touch.

On my last morning in Cairo, my new HU connection insisted that I join him and a few of his friends on a leisurely breakfast cruise along the Nile in a traditional white-sailed felucca. I could think of worse ways to spend a morning than taking a slow boat along the Nile.

By mid-afternoon, I was back on the bike and battling my way through the crazy Cairo traffic. It took well over an hour to clear the city, and I was soon rolling into the desert, past the pyramids, and away from the chaos.

CHAPTER TWENTY-ONE

Lonely Outpost in the Desert

Personal Diary Entry: 13th February 2014

Day 276 (of 549)

39,721 km of 62,840 km (24,681 mi of 39,047 mi)

As I sit here writing under a hot desert sun, the ghostly white strip of skin where my wedding ring once sat is slowly beginning to disappear, like the feelings I once had for the person who placed that ring on my finger. With this comes the realisation that for the first time in many years, I am truly free to follow my own path.

It wasn't hard to resume the trip south again. After almost a week in one of the world's busiest cities, the prospect of a solo adventure in the peaceful desert was most appealing. I knew I'd been distracting myself, and I needed more time to address my fragile emotional state. I had to be alone to do so. That is precisely why I picked the Oasis Road.

Four potential routes were possible from Cairo to my eventual destination, Aswan. After talking to local riders, I found out that the Oasis Road took a long, lonely detour through the Western Desert and was reportedly the least travelled of all the southerly routes in Egypt.

I often had the road all to myself.

As the kilometres rolled by and I got deeper into the desert, I could feel a sublime peace settling over my mind and body. The desert had a peculiar energy to it, something entirely alien to an Irishman. There was a sense of serenity, ominously underscored by a subtle danger. As beautiful as it was, it could also be deadly, and that unique combination inspired a feeling of wonderment.

With fuel in short supply, I filled my auxiliary tanks for the first time to test the true range of the bike while fully loaded. I had called ahead to my first stop at Bahariya Oasis, over 400 kilometres from Cairo, and they told me they had not had any fuel delivered for several days and the entire town was dry. The next possibility for picking up fuel would be another 200 kilometres beyond that, at Farafra Oasis, and they were unsure whether or not there would be any fuel. From there, I wouldn't find fuel for another 200 kilometres beyond Farafra, which I knew was well beyond my maximum range.

I set off anyway, knowing it would all work out.

Besides, I could enjoy the desert more, even if I had to spend a few extra days waiting for a fuel delivery. I was constantly surprised at how the endless dunes seduced me with their austere beauty and chilling indifference. I felt an intense sensation of belonging amid the parched landscape. The solitude brought me peace, and the silence gave me time to think and process my grief without distraction. Its vastness demanded respect and humility, yet it also bestowed a unique sense of awe and wonder.

My late departure from Cairo left me on the road with not enough time to beat the sunset, and after my first experience in Egypt of driving at night, I had no desire to have another. With the gear I carried on the bike, I could stop and camp anywhere, so I told myself if it got too sketchy, I would just pull over. The lack of traffic and the reasonable road surface allowed me to make good progress, and by the time it was completely dark, I knew I was only within an hour of my intended stopping point. So I pushed through.

It may not sound like such a big deal, but driving at night in Egypt could be a white-knuckle, butt-clenching experience. The local habit of driving without lights still baffled me. More often than I expected, I encountered the unnerving sight of the bloated corpse of a camel on the side of the road.

And then there were the potholes.

These nocturnal demons liked to come out at night, especially the big ones. They were hard to spot until I was almost upon them, and then I was faced with the dilemma of whether or not to weave to avoid them on a questionable road surface or gun the throttle to lift the front wheel over them. Thankfully, I made it to the first oasis without any serious mishaps.

Soon, I was settling into a quiet lodge belonging to a wise old Bedouin named Mahdi, who had seen many overland bikers, so he invited me to park my machine inside the lobby of his humble residence. Within minutes, he had a hot meal waiting for me after my long ride. Then he showed me to my comfortable room. After a great night's rest and a delicious, traditional breakfast of eggs, beans, and bread, we sat talking in the sunny courtyard about the road ahead and the roads we had travelled.

"So, how do you like the Desert Road, my friend?" asked Mahdi.

"I like it a lot. It's so . . . serene," I replied.

"Yes, it can be peaceful but also very dangerous, so you must take care. Sometimes the sandstorms can turn day into night," he informed me. His warning sounded ominous, but my mind struggled to imagine a storm so fierce. I acknowledged his caution and instead focused on what I considered more relevant.

"Can you tell me more about the road ahead?" I asked.

"From here, you will pass through the Black Desert and then the White. You will not see many other vehicles so if you have trouble, you may have to wait for a very long time before help comes. Make sure you have a lot of water. Some villages you pass will not always have fuel, so add more to your tanks every chance you get. Do not pass gas."

He chuckled at his own joke.

Then he rose and shuffled over to a nearby table, returning with a handful of business cards. He placed them on the table and slid them towards me as he sat down.

"If you meet travellers coming this direction, please give them one of my cards and tell them about my place." He smiled, then added, "Tell them about your good friend, Mahdi."

He sat there, looking at me expectantly. It took me a second to realise what he wanted.

I fumbled through the pockets of my heavy motorcycle jacket and pulled out my wallet. Inside it, I still kept a few of the contact cards Sarah and I had made for occasions just like this. On the card, alongside our contact details, was a small photo of us together. My heart ached, and my stomach churned as I looked at it.

"What is wrong, my friend?" Mahdi asked.

I hesitated before handing him the card, and then I watched his expression as he studied the picture. Eventually, his eyes met mine, and I felt he could sense my pain.

"Where is this woman now?" he asked softly.

"She left me." My voice caught on the words as I struggled to compose myself. "I lost her in Israel."

I told him the story of what had happened, pausing occasionally to collect my thoughts and control my emotions. It still felt so raw, even though I had been trying to block it all out of my mind for some time. Talking about it stirred up some powerful feelings that I had hoped to bury, but I knew I couldn't hide from them forever. When I finished talking, I realised I'd become extremely tense. I slumped back into my chair and exhaled a long sigh. We sat silently for a while. I hadn't realised until then just how quiet it was at this lonely outpost in the desert.

"Fucking Israelis!" he spat, breaking the silence. "Always taking what does not belong to them."

I was temporarily stunned by the mixture of sadness and anger in his voice. His gaze drifted to some unseen event, and I wondered what distant trauma he was reliving.

I chose not to enquire.

I hadn't noticed his frailty until then, and as he rose from his chair, he appeared to have visibly aged. He extended his hand, and I stood to receive it, his grip still firm and reassuring. He looked at me intently with a sincere smile.

"My friend, I wish you every success on your journey," he said, "I am sure you will find all the answers you seek, *inshallah.*"

***

With no fuel available in the Bahariya Oasis, I topped up my main tank from my external auxiliary supplies, totalling an extra nine litres, just over two gallons. I hoped that my next stop in Farafra would have more.

The road south took me through the Black Desert, where the sand was coated with a dark, thin layer of volcanic dolerite from the surrounding mountains. This gave the impression of riding across the surface of a dark, apocalyptic wasteland.

It suited my emotional state.

Further south, the road cut through the edge of the White Desert, where wind-sculpted, snow-white rock formations created an even more surreal landscape. As I entered the region, I rode into the heart of a powerful sandstorm. Mahdi was right. With little hope of finding shelter, I pushed on. The visibility dropped and sand penetrated everything. When I could not see more than five metres ahead, I adjusted my speed accordingly.

Suddenly, I felt a sharp sting piercing my right earlobe as though something was biting me inside my crash helmet. I flinched and shook my head to dislodge whatever creature had crawled inside, my imagination conjuring images of all kinds of scorpions, spiders, and scorpion-spider hybrids. Then, another sting pierced my left earlobe, stronger than the first.

Buffeted and sandblasted by the powerful crosswinds, it took me a moment to realise it was an electrostatic shock from the windborne particles colliding with my helmet. It was a relief to know it wasn't an unwanted passenger, but the shocks continued as the sandstorm raged on. Coated in a fine white powder, I eventually emerged on the south side of the maelstrom, crunching sand between my teeth.

I felt a little raw after several days of involuntary exfoliation in the desert.

The road across the final desert plateau had taken much longer than I anticipated, the surface deteriorating with every kilometre I travelled. Cracked and twisted by the intense summer heat, when it disappeared altogether, I had to pick my way between large mounds of sand to find a way through.

After that, the road became a muddy, slippery mess for many kilometres as it was clearly still under construction. I was thankful for the semi-off-road tyres I had fitted to the bike several months earlier. The atrocious conditions caused me to be on the road to Luxor after dark yet again. As I dropped into the fertile Nile Valley, it became apparent that my arrival had coincided with the annual sugar cane harvest.

Overloaded tractors and trailers were piled high with sugary stalks, forming long convoys of what looked like gigantic snakes of vegetation. Children lined the roadside, chewing on the sweet bounty of stray cane that fell from the precarious loads.

Luxor proved to be a fascinating city on a beautiful stretch of the Nile River with an extraordinary concentration of ancient ruins. I couldn't resist exploring a few local sites and spent a day longer than planned in the area. With Aswan so close and the ferry to Sudan departing only once a week on a Sunday, I eventually pushed south to give myself enough time to organise the paperwork and finalise booking arrangements.

Though relatively short, the road from Luxor to Aswan was bustling, and every town along the way had installed some of the most brutal speed bumps I'd ever experienced. For the final 200 kilometres, I rarely made it out of third gear.

It took me four hours to complete a two-hour stretch.

I eventually found my way into Aswan, Egypt's southernmost city, and, in character, more African than anything I had experienced. The Nubian culture was more dominant here, and the pace of life much more relaxed than in the north.

I had planned to load my bike onto a slow barge to Wadi Halfa in Sudan on a Saturday. The following day, I hoped to begin the eighteen-hour journey by passenger ferry across Lake Nasser. A road was currently under construction that would eventually circumvent the need for the ferry, but it was not yet fully open to the public. I prepared myself for a long, relaxing boat trip, resting my body after a rough ride and finding the time to poke around in my brain a little.

Egypt was the first country on this journey where I had to travel without Sarah. Many times, I was alone, but I never felt truly lonely. The people I met along the way were wonderful; I doubt I would have received as much kindness had I not travelled solo.

As I prepared to move on, I looked back at how the trip had evolved and focused on the positive experiences that may not have occurred had things turned out differently. With that came the opportunity to practise forgiveness, to liberate me from the anger and bitterness that darkened my thoughts for so long – those ugly, primal emotions that fed into negativity and depression.

That was not how I wanted to proceed.

CHAPTER TWENTY-TWO

Erase My Footprints

Personal Diary Entry: 20th February 2014

Day 283 (of 549)

40,461 km of 62,840 km (25,141 mi of 39,047 mi)

There are still ups and downs, but I take each day as it comes and realise that it is my attitude that will define how I perceive the world and how the world will perceive me.

The Nile River, the world's longest watercourse, had been my travelling companion since leaving Cairo. I never strayed too far from its lush, verdant valley as it cut a winding, green path through the heart of the Sahara Desert.

The river's waters have supported life in the region for thousands of years, and it was lined with settlements old and new as countless civilizations have risen and fallen upon its banks. The contrast in terrain just a few kilometres from the river was truly remarkable, from fertile farmland to barren desert.

As hostile and uninviting as it looked, the desert fascinated me.

It was hard not to think of the nomadic people who lived within, the Tuareg and the Bedouin among them. On several occasions, I deliberately wanted to get lost. I'd leave my bike behind and wander into the dunes until it was out of sight. I would close my eyes and spin around until I was unsure which way led back to the road before sitting down to meditate, allowing the wind to slowly erase my footprints. Only then would I feel fully immersed, my senses heightened by the danger I had put myself in.

On one occasion, when I opened my eyes, a young Bedouin boy stood nearby, quietly watching me. I greeted him in Arabic, and he nodded and smiled in return. I had not heard him approach, and I was uncertain where he had come from. I was sure I was a long way from any settlements. I swept my gaze along the horizon, scanning for signs of a caravan or a cluster of tents, but there was nothing. When I looked back to where the boy had stood, he was gone. He had vanished so silently, so completely, that I questioned whether he had been there at all.

***

Fixers are guides, or local experts, who are familiar with the local customs and practices, well-versed in relevant issues, who sometimes also serve as translators. They usually work behind the scenes and are often paid in cash.[15]

Although I had hoped not to use fixers for border crossings, managing the logistics of getting through the Egyptian/Sudanese frontier and crossing over a 400-kilometre portion of Lake Nasser by myself were two stressful tasks that were rather complex.

I began to realise I would need help.

A friend in Cairo had recommended I get in touch with a local acquaintance, Kamal, should I have any difficulties. While I'd hoped I wouldn't need the connection, within hours of my arrival in Aswan, I encountered several significant problems.

According to the ferry company, all seats were fully booked on the boat I had intended to take, and the next alternative would leave in seven days. Even then, it was not guaranteed the next sailing would coincide with a suitable barge departure that could carry my motorcycle. Ideally, I travelled with my bike at all times, but in this case, it was simply not going to be possible.

I gave Kamal the Fixer a call.

In less than an hour, he arranged to get me the sailings I wanted. I set up a meeting with him the following morning to discuss the details, and when we met, he struck me as a very relaxed and competent person I could work with. We agreed on his fee, and for the next two days, we worked together to clear the bike through the exit procedures and then get my paperwork in order.

With Kamal's help, acquiring the Sudanese visa was relatively straightforward. With an office in Aswan that charged half the price of the Cairo branch and took a fraction of the time, I was soon stamped up and ready to go, the visa being a prerequisite for purchasing the ferry ticket. A guidebook recommended allowing up to six weeks if applying for the same visa through the Sudanese embassy in London. I had mine in under an hour.

Kamal and I took the motorcycle to the port after I had stripped a few essentials off it to keep me going while the bike was in transit. It was often a slow process to obtain the right authorisations in the correct order, but Kamal knew the procedures intimately, so it took only a couple of hours.

Oh, how the Egyptians loved their paperwork.

It seemed as though he was on first-name terms with every official at the port, and, at one point, he was even pushing a few of the soldiers around in a jovial manner. After removing the Egyptian licence plates and clearing the Carnet de Passages, we were soon lashing the motorcycle to the deck of a rusty old barge bound for Wadi Halfa, set to leave the following day.

On the day we loaded the bike, the port was virtually deserted, but when we returned the following afternoon to secure my passage on the ferry, the place was a hive of activity; porters struggling with oversized loads, ticket agents yelling at each other, a handful of sweating westerners in the chaotic line hoping for a few last-minute cancellations, the police and army strutting around looking officious. Among all the hustle and bustle, here and there, an individual would slowly spread out a prayer mat and quietly go about their devotions, oblivious to all the noise and mayhem.

We had arrived at the port shortly after noon, and on my behalf, Kamal spoke to the infamous Mr. Salah, the general manager of the last company to service this route. Mr. Salah appeared to relish his power as he wandered nonchalantly through the heaving crowds, smartly dressed, eyes hidden behind mirrored aviator sunglasses. He had recently refused passage to a European cyclist who was impertinent enough to enquire as to why his fare differed so much from the amount paid by locals. With my motorcycle already loaded on another boat and the ferry reportedly full, I was a little anxious at the prospect of my bike sitting in Sudan unattended for several days if I could not catch this sailing.

With Kamal's connections, I secured one of the last three available seats. I felt somewhat guilty knowing the people still queuing outside the ticket office were probably out of luck. By 3 p.m., I was on board, and with the departure scheduled for 6 p.m., I still had plenty of time to settle in. I watched in wonder as the boat crew slowly packed the vessel to full capacity and then well beyond – even the lifeboats filled with people looking for a spot to stretch out and get comfortable for the eighteen-hour voyage. Passengers and packages occupied every square inch.

As I settled into my assigned area, I wasn't sure what was going on. Kamal had either performed a miracle or told people I had some kind of rare disease – everyone gave me a wide berth. After being wedged between too many people in big cities and long ferry boat lines, I was happy to be able to spread out a bit. I stretched my limbs and balled my motorcycle jacket into a lumpy pillow.

An hour later than scheduled, after some delays with loading the boat, finally, the gangplanks were lifted, the tethers released, and we were on our way. The same rules seemed to apply here as with Egyptian traffic – the boat moved through the night without any lights other than the ghostly glow from the instrument panel inside the wheelhouse. As soon as we cleared the port and all background light faded into the distance, the evening sky came to life with a riotous display of stars I hadn't seen since my time in the desert. When the partial moon dropped below the horizon, even more stars appeared, and the Milky

Way splashed high and wide overhead. I spent most of the night lying on the hard steel deck, gazing at the free light show, picking up the hint of new constellations on the southern horizon as we neared the Tropic of Cancer.

Ancient travellers used the heavens to guide their way, so they named clusters of stars and lines of latitude to help them navigate. The Tropic of Cancer is an imaginary line that marks the northern boundary of the tropics, at 23.5 degrees north of Earth's equator. When the Tropic of Cancer was first named about 2,000 years ago, the vertical rays from the sun struck in the direction of the constellation of Cancer, causing the northern hemisphere to receive more direct sunlight during the June solstice.[16]

But things change.

Modern travellers can access GPS devices or a simple Maps app installed on their phones. The Tropic of Cancer no longer resides in the constellation of Cancer, although it hasn't been renamed. And things are going to continue to change. The Tropic of Cancer is expected to move to new latitudes as Earth's tilt causes the location of the sun to change.[17]

If Earth's obliquity can change, so can people's feelings for each other.

I thought of Sarah's feelings for me changing – was it over time or all of a sudden? I thought of how I no longer missed her every single moment of every single day. I thought of how utterly broken I felt when we first went our separate ways and how capable of healing I discovered I was after some time had passed.

***

Above deck, aboard the ageing ferry, a light breeze kept the bugs at bay, and it was pleasant enough to sleep with the motorcycle jacket changed from a pillow to a cover for warmth. Occasionally, I would get up and wander below deck to stretch my legs or visit the overflowing toilets amid the stale, dank air within the packed vessel. It was uncomfortably warm inside, and I was thankful for the space I had out in the open air. I was able to sneak in a few hours of sleep, but at one point I was awoken by something heavy scampering across my chest on top of my jacket. I didn't get a chance to see what it was, and I hoped it was only a cat – even though it did seem a little too light.

Shortly before dawn, I gave up on getting any more sleep, so I readied myself to catch the sunrise on the portside of the boat. In the early morning light, as I looked out across the ocean-like reservoir, I saw solitary fishermen hauling their nets onboard small wooden boats as hungry pelicans patiently watched nearby. And still, we moved south, gliding past the relocated ruins of Abu Simbel and into Sudanese waters. At noon, the distant port of Wadi Halfa came into view, and the restless passengers prepared themselves for the chaos ahead.

By one o'clock, they berthed the ship, opened a tiny door, and the exodus began through the tight bottleneck. Travelling light, I was able to get off the boat with ease and was soon walking along the jetty towards the customs inspector, a smile on my face and a new country under my feet. After a cursory glance at my passport and a few simple questions from a border official, I was through the gates and on my way to the small village of Wadi Halfa.

Kamal had called ahead, so his friend was waiting outside the port gates to give me a ride to a cheap hotel nearby, where Kamal had thoughtfully reserved a room for me. He had told me it would be quite tricky to find accommodation on the day the ferry arrived, as the sudden influx of people often overwhelmed the few guest houses.

Kamal's friend, also a Fixer, arranged transport for overlanders travelling north. He filled me in on the procedures required for entering Sudan. Already, he had several other motorcycles and a few trucks ready to load for the return crossing to Aswan. He also told me that engine problems had delayed the barge carrying my bike, so I should expect to stay in Wadi Halfa for a few extra days.

I had never imagined the 400 kilometres separating Aswan and Wadi Halfa would bring with it such a noticeable change. Crossing the Red Sea had been a major milestone on this voyage, but sailing to the southern side of Lake Nasser felt like I was immersing myself in a different world altogether.

In north Sudan, I felt like I had truly entered Africa.

Although still predominantly Arabic in their customs and beliefs, there was a significant contrast to what I had experienced until then. The inescapable raw power of the landscape reflected in the resilience and vigour of its inhabitants. I experienced a generosity that belied the obvious poverty. On many occasions, total strangers would insist on paying for a cup of tea or a bag of bread rolls. It was only pennies, but in each case, it really was the thought that counted. The locals took a noticeable pride in how they presented themselves and how outsiders perceived them.

Before leaving Israel, I had been advised "not to trust the Arabs," ironically, by a deceitful bastard who turned out to be one of the least trustworthy men I had ever met.

Instead, I discovered an honour among those I met there.

There were many times when I had no choice but to leave my bike to purchase food or find a place to stay, and it could be stressful knowing that if a person wanted to, they could easily have made off with some or all of my gear. Obviously, I used some common sense when choosing where to leave it, and I locked most of the equipment to the frame, but that would only have slightly discouraged the opportunist. A technique I'd picked up from my Australian friend in Jordan had served me well since. If I needed to leave my bike, I would choose someone sitting nearby and ask if they thought my stuff would be safe while I was away. If they answered "yes," they tacitly assumed some responsibility, and when I returned, I would thank them graciously before moving on. If someone was determined enough, they could have easily gotten past my basic locks. What took me aback was that no one ever tried.

Sudan was full of surprises.

The main roads were in excellent condition. Some of the revenue from their lucrative oil industry was put to good use improving infrastructure. Many Chinese civil engineering companies were involved with the expertise and human resources to complete major projects on time and within budget. Moving south from Wadi Halfa, I was initially riding through a harsh, unforgiving desert, but gradually I noticed more foliage alongside the road. Finally, it felt like I was nearing the edge of the Sahara Desert.

I was already two months behind schedule.

I had hoped to avoid some of the strong winds that ravaged the region at that time of year. Even though the sandstorms added some excitement to the overall experience, their ferocity was expected to increase, and the thought of being caught in one that "can turn day into night" filled me with dread. The daytime temperatures were pleasant and the air was dry, so the heavy motorcycle apparel I constantly wore didn't feel too oppressive.

I was the only foreigner in many towns and villages I stopped in. Within minutes of arriving, a crowd often gathered around my bike, keeping a respectful distance but always asking the same initial three questions:

"How fast?"

"How many CCs?"

"How much does it cost?"

Ancient temples were abundant, often close to the river and easily accessible from the road. The pyramids in this region were from the period of the Meroitic Pharaohs, with a more elegant, slender appearance than their northern counterparts. I had every site all to myself – so few tourists ventured into this part of Africa. One day, I would have been glad for some company when I buried the bike up to its axles in the deep, soft sand. It took a combination of ingenuity, brute strength, and some choice swear words to get it back onto the firm surface of a dirt road.

On the evening of my first night in the sleepy village of Dongola, I was invited to attend a festival nearby. I was greeted by the loudest music I had heard in a long time and hundreds of men in traditional long, white robes brandishing swords and sticks, and dancing in a style reminiscent of the Whirling Dervishes of Turkey. Sudan was operating under Sharia law, with alcohol prohibited, but the atmosphere was intoxicating as they spun feverishly in the warm evening air. The only women to be found were gathered in quiet groups around the edges of the party.

One aspect of the local culture I found hardest to accept was the treatment of women. Openly regarded as second-class citizens and heavily swathed in their *hijabs* and *chadurs*, they often appeared solemn yet sorrowful. Female genital mutilation (FGM), was widely practised in Sudan, with an estimated 85% of women having undergone some form of the barbaric procedure.[18]

I spoke with an OB/GYN doctor from the local hospital in Dongola who had countless stories of how this practice had ruined the lives of so many. Any culture or creed that deems it necessary to interfere with the bodies of their children in this way needs to be questioned. I have heard many arguments for and against such procedures, but I wondered, why not let the individual involved hear those same arguments so they can make informed decisions? FGM was often mistakenly associated with Islam, but there are no religious texts that support or require the practice. It was deeply rooted in the cultural traditions of this society and supported by a lack of education and awareness, mostly in rural areas.

Only on my arrival in Khartoum did I first notice women openly smiling.

Khartoum is not one city, but three, clustered around the confluence of the Blue and White Nile Rivers. It struck me as surprisingly modern, compared to the rest of Sudan, with glass-tower blocks stretching into the skyline and a busy airport close to the centre. I had decided to stay at its one and only youth hostel. Other travellers I had met described it as cheap and convenient, with plentiful parking within its open compound. There were even rumours of possibly getting a hot shower.

With my crude maps and trusty compass, I eventually stumbled across the hostel. Street signs were rare here, and, as with much of the Middle East, locals gave directions in relation to nearby landmarks. The other travellers were right; it offered secure parking and comfortable dorms, which was everything I needed, so I decided to make it home for a few days – even though the showers were freezing cold.

With the only crossing into Egypt so close, I was beginning to encounter more overland travellers as they converged upon Wadi Halfa and the ferry to Aswan. Most were moving north, so it was good to pick up tips and advice for the regions I would travel through, for road conditions, places to stay, where to eat, and areas to avoid.

All were couples, which often made me wonder how different that part of the journey could have been for me, but I did not dwell on those thoughts for long. Those travellers were brimming with stories of adventure and adversity as they came to the end of their journeys, and it filled me with excitement to listen and wonder about what was in store for me. One recommendation they often repeated was using discretion when taking pictures. It seemed as though the local security services were suspicious of foreigners who took too many photographs.

With over 40,000 kilometres – roughly the circumference of Earth – behind me and an untold number ahead, I felt like my journey had entered a new phase. Africa felt wilder and less predictable than what had come before. Conditions were considerably more challenging and constantly assaulting my senses. It was somewhat overwhelming – how different everything was visually, culturally, and in every other way. There wasn't much time for travelling down memory lane as I had to adapt to the ever-changing present.

It was exactly what I needed.

CHAPTER TWENTY-THREE

Riding into the Unknown

Personal Diary Entry: 6th March 2014

Day 297 (of 549)

42,511 km of 62,840 km (26,415 mi of 39,047 mi)

The sound of an engine is rare in Lalibela. Days begin slowly with the crowing of roosters and a cacophony of delightful birdsong. Nights are accompanied by the chirping of crickets and the occasional barking of dogs.

The people are friendly – willing to talk, and quick to smile. There are children everywhere, marching to and from school or playing with homemade toys in the dusty streets, always asking for pens or pennies from the tourists who wander by.

Time has slowed down. Jobs get done when those who do them are ready – no sooner, no later. The pace of life is so relaxed, it is contagious. My original plan to stay for two nights has now been stretched to three, and I still haven't decided if I will move on tomorrow.

I spent several days more than I had planned to in Khartoum. With relative ease, I was able to obtain my Ethiopian visa, but that didn't mean I was ready to move on right away. Instead, I explored some of the large markets around the edge of the city, where every kind of product I could imagine was on display in the stalls that lined the narrow streets and filled the adjoining warehouses. In the corner of one large building, I stood in awe at the foot of a literal mountain of used guitars.

A couple of overlanders from Greece, coming to the end of their two-year African odyssey, made for good company and a perfect opportunity to pick up a few tips about some of the regions I would be travelling through. On a hot afternoon, as I strolled across a bridge into the older part of the city with my new Greek friends, we all felt quite dehydrated. When a truck carrying bottled water drove by, I gestured to the driver that a bottle would be nice. He unexpectedly stopped while his passenger ran across the busy road to give each of us a free sample.

The kindness of the people of Sudan constantly touched me.

With only a few days in a city, getting a feel for the layout was often confusing, so asking a taxi driver could sometimes be the easiest way to get directions. Typically, in other cities, they would insist you get on board so they could take you there, but in Khartoum, they would help you as best they could and expect nothing in return. Unfortunately, there were times when some of the directions would be questionable, but they would rather tell you something than let you down.

Soon, it was time to move on, and I decided to follow the Blue Nile towards the romantic city of Wad Madani, Sudan's most popular honeymoon destination. During the daytime, it was a hot and dusty riverside market town, but as soon as the sun set, the area along the riverfront came alive with couples and young families enjoying the warm evenings and the languid Nile. I took a cheap room at the Continental Hotel, a crumbling colonial relic with large, foliage-filled gardens and tired, wicker furniture. I spent the evening on the patio, watching the townsfolk stroll by in their finest attire.

The border to Ethiopia was tantalisingly close, so I rose early the next day. I prepared for a drive that got me closer to exiting Sudan, but I knew I wouldn't make it all the way out of the country in one day, especially with how long some border crossings took.

As I rode into the town of El-Gadarif in the late afternoon, I passed through the outer suburbs, where I noticed a young boy standing by the roadside. When I got closer, he casually reached down and picked a bottle off the ground, and cocked his arm behind his head as though he was about to throw it. I had seen this behaviour many times since entering the Middle East, but it rarely led to anything – most likely, those mischievous rascals just wanted to see me flinch. But on this occasion, he followed through and launched the bottle in my direction, which took me completely by surprise. Only after the bottle bounced harmlessly off my leg did I realise what he'd done.

I had always wondered what I would do in such a situation but never figured out the best response ahead of it happening. I knew I needed to communicate that the young boy's behaviour was unacceptable and all actions have consequences, so I turned the bike around and rode back towards him. As soon as he saw me turn, he ran off over an expanse of rugged wasteland towards a nearby market, assuming I wouldn't follow. I chased him for a short distance until he disappeared between the tightly packed market stalls. Perhaps he would reconsider his behaviour the next time he had the urge to throw anything at a motorcycle.

Or maybe I'd just helped to start a new sport.

Disregarding this unusual welcome, I decided to stop and spend my last night in Sudan in this busy market town. El-Gadarif lacked the charm of Wad Madani, but it was interesting to wander through the bustling markets and marvel at the ingenuity on display.

When discussing recycling in the West, we often assume we are doing our best by separating our waste and putting it in the correct bins. In Sudan, recycling meant breaking down used goods into their most basic components and finding a use for everything. Stalls full of what we would consider junk lined the market, and people were hard at work repairing everything they could. Watches, shoes, mobile phones, umbrellas, and bicycles were among the items we would have long since thrown into a landfill in the West. I found one section of the market occupied entirely by tailors, so I had a few sturdy repairs made to some of my battered clothing.

Even after dark, the town felt perfectly safe.

Once the markets had closed for the day, I strolled through the darkened streets, looking for a suitable place to eat. Simple stalls selling all kinds of food came to life in the evening, and soon, the streets were buzzing with people eating and relaxing.

Alcohol was outlawed, possession of which was punishable with up to forty lashes, so public intoxication was almost unheard of. Tea was the drink of choice, served steaming hot and impossibly sweet. I would get strange looks when I asked for tea without sugar, which they often interpreted as just two heaped spoonfuls instead of the mandatory four.

The staple dish of Sudan, found virtually everywhere, was *fūl,* pronounced "foul." Made from stewed fava beans, oil, and spices, it often varied in consistency and quality and occasionally lived up to its name.

When eating questionable food during this trip, it had become a common practice of mine to wash it all down with a magical potion I had discovered in Africa. Whether it was the phosphoric acid, the caffeine, or simply the placebo effect, Coca-Cola seemed to kill any undesirable bacteria in my stomach. I wasn't particularly fond of the taste, but it was widely available and often served me well throughout my journey. Against all the odds, I never once had a single stomach complaint.

***

After a peaceful night in El-Gadarif, it took a little over an hour to reach the Sudan-Ethiopia border. I almost rode past the final checkpoint on the Sudanese side without the necessary stamps. Helpful locals directed me back towards the various unmarked shacks where I could clear security, customs, and, finally, immigration. Of course, it was tea time at the customs compound, and everything was on hold, so the officials invited me to join them for a sweet, refreshing cup under the hot morning sun. Before long, I was allowed to enter no-man's land and cross the small bridge into yet another frontier.

Entering Ethiopia was reminiscent of Dorothy's arrival in the Land of Oz – just without the munchkins. Everything suddenly appeared to be in vivid technicolour. Even the national flag's red, gold, and green – with a big golden star in the centre – seemed flashier than Sudan's simpler version. Plus, gone were the chador and hijab. After the rather drab, conservative dress codes in Sudan, Ethiopia was bursting with radiant colours and provocative styles.

I'd almost forgotten what the female body looked like!

In Ethiopia, some women's outfits on display left little to the imagination. Risqué Western fashions blended with vibrant African patterns to form unique creations, and accompanied by a general cheerfulness, even among the border officials, I immediately felt eminently welcome.

I was soon riding into the nearby city of Gondar, excited about this joyful, new atmosphere. A couple of other travellers I'd met on the road suggested I stay at the Belegez Pension, near the centre of town. It promised secure parking, fair prices, and even the chance of hot, running water and electricity – although not necessarily all at once.

On my first attempt to find my accommodation, I immediately got lost.

As I looped around the seventeenth-century castle that dominated the town centre for the third time, a young man flagged me down and asked where I was planning to stay. He sent me off in the right direction and even turned up shortly afterwards to make sure I'd found the place. I got talking to him and learned I had arrived at a rather auspicious time.

Turns out, a two-month-long period of fasting was due to begin the following morning, so that evening, the town would be in full party mode before abstaining from alcohol, sex, and animal products for the fast. My new friend promised to stop by later that evening to show me where the locals hang out. Meanwhile, I explored this thrilling new city on my own.

The Ethiopian people liked to do things their way.

Depending on who I asked, the date was sometime in June 2006, which meant I would turn thirty-five again soon – even though my calendar said it was February 2014. Time appeared to be measured in relation to the sun regardless of the time of year.[19] It took a little getting used to, but it seemed to make sense.

The food was mouthwatering, spicy, and bursting with flavour and colour, often served atop a large, rubbery pancake called *injera,* which replaced the need for plates and utensils. The sour-tasting injera seemed a bit odd to me at first, but it complimented the savoury dishes well. To wash it all down, they had delicious beer served up at room temperature, so I did have to specify that I'd like it cold.

As the original home of coffee, the drink was available everywhere. Often, ordering a cup involved an intricate ceremony where the beans were roasted, ground, and passed around for inspection while incense burned and water slowly boiled over a small charcoal fire. Eventually, they presented me with an espresso-size cup of the best coffee I had ever tasted, with a bold, complex aroma and a rich, exquisite flavour. I'd never been a huge fan of coffee, but after having it in Ethiopia, I felt everything else would pale in comparison.

After eating and drinking my fill, my new friend, Moulish, who had directed me to my accommodation, returned to see if I wanted to join him for a night out. We began our evening at a sleepy little hotel in the centre of town. I was beginning to think it would be a quiet night until an attractive young woman approached me and persuaded us to let her join us.

After learning the absolute basics of Arabic, I was now fully out of my depth, yet again, with the local language of *Amharic.* Unlike anything I had heard or read before, it seemed to save some of its longest words for the simplest of greetings. So I reverted to the tried and trusted method of talking slowly and loudly in English, waving my arms about and using the "old dog, new tricks" excuse for my linguistic laziness.

As luck would have it, my new acquaintance, who went by just "Mary," spoke very little English – so we got on quite well. So well, in fact, that she insisted I make a solemn promise to sleep with her before the night was over.

At least, that's what I think she said.

I was suddenly reminded of an old Groucho Marx quote about not wanting to be a member of any club that would accept people like me as a member, so I made my excuses and left the bar.[20] Moulish joined me, and we ended up at a local's club where the atmosphere couldn't have been more different. The club was packed, humid, and hot, and the music was loud. A remarkably talented selection of locals took to the stage and entertained the audience, dancing in styles reflecting the regional variations in each of the chosen tunes.

Had I not been with Moulish, I'd have barely noticed the subtle differences, but all the variations included the "shoulder dance," an impossibly fast shaking of the upper body that, even when perfectly in time with the music, could have resembled a cross between a demonic possession and an epileptic seizure. It was remarkable to watch but impossible to imitate, so I sat quietly on the edge of the dance floor like an over-excited teenager, eager to join in but too self-conscious about making a fool of myself. The atmosphere was electric, and the night stretched well into the early hours.

The next day, I checked out the old castle complex in the town centre and then raced to the nearby hilltops for evening cocktails as the sun went down. There, standing before a pastel-hued panoramic view of the ancient city of castles with some unexpected modern structures mixed in, I realised I could have easily pushed the pause button on this trip and stayed indefinitely in Gondar. The prospect of gorging myself on all this new country had to offer was too enticing to resist, so I knew I had to force myself to continue on before it drew me further in. After a few more days, I loaded up and turned north towards the Simien Mountains.

The road from Gondar to Aksum turned out to be another gem.

Still under construction, the road offered a long day of very mixed conditions, from perfect, silky smooth asphalt on the completed sections to rugged dirt to talcum-powder-fine sand so dry it squeaked underfoot, all through some of the most spectacular mountain scenery with breathtaking views at every hairpin bend. I kept thinking how, when they finished building, that road could become one of the finest rides in the world, provided the heavy traffic didn't tear it up before then.

As I had done very little research before entering this part of Ethiopia, I hadn't expected it to be so mountainous. But what I found delighted me, and the bike never ceased to impress with how well it handled the varied terrain. As night fell, I rolled into Aksum, exhausted and exhilarated by the long day's ride.

The town of Aksum was once the home of the Queen of Sheba.

That's what many said, at least, and it was easy to see why. One of the most important ancient sites in all of sub-Saharan Africa, Aksum was dotted with the relics of an illustrious past. Delicate obelisks peppered the area, some as tall as thirty-three metres. Sixth- and seventh-century tombs of long-deceased nobility were everywhere, and with, according to local legend, vast quantities of treasure still concealed within.

By far, the most significant feature of the town was the carefully guarded Church of St. Mary of Zion, believed by many Ethiopians to contain the original Ark of the Covenant. Foreigners were kept well away from this holiest of shrines, with the well-hidden Ark protected by a single monk. It is said that all others who dare to gaze upon it would be immediately struck dead. When I told a local guide I was willing to take the chance, he looked at me with absolute horror.

Their faith runs deep in that country.

***

On one particularly frustrating afternoon, I had exhausted all local options in my search for fuel. After the last place I thought of in Aksum turned me away, a young boy approached me who promised he knew where to find it. He suggested he hop onto the rear seat of the bike and direct me.

"Why not?" I thought as I nodded. He gave me directions by tapping my shoulders. After we checked a few more places in his hometown, he proposed we venture to a neighbouring town. I strapped my crash helmet onto his head and drove us over thirty kilometres from his home.

We succeeded – I found fuel! The boy had the biggest smile on his face as he again clambered onto the rear seat.

"I must tell you something . . . " he began.

"Yes, go ahead," I said to encourage him.

"This was further and faster than I ever travelled before!" His voice was quivering from excitement.

As I revved the engine, I knew just what I had to do.

On the return journey, I pushed the bike over 180 kmph while he screamed with joy, wrapping his arms tightly around my waist. He asked me to drop him outside his school when we returned to his town, as he couldn't wait to tell his friends about where he had been and how fast he had gone.

From Aksum, I turned east and was soon enjoying the amazing roads of the mountainous Tigray region, stopping for a few hours to explore the Debre Damo monastery. Perched on a sheer-sided, flat-topped mountain, I could only access it by hauling myself up a twenty-metre cliff face on a weathered-looking, old leather rope.

After a stopover in the busy university town of Mekele, I resumed a southerly route, up and over steep mountain passes, through small villages and bigger market towns, avoiding the menagerie of animals that would occupy the road. Monkeys, camels, donkeys, sheep, goats, and snakes all helped to keep me focused as I put the kilometres behind me.

Then I began to run low on fuel – again.

Every service station I stopped at had run out of petrol, and no eager young boy appeared to help me find it. After asking around, it turned out it was not that difficult to find fuel on the black market, but it was expensive and the quality was questionable. There was always someone nearby willing to sell individual litres of black-market *benzine* for twice the regular price and, sometimes, half the purity. I carried a useful funnel to filter out debris and water, which proved to be a priceless part of my kit.

Every time I had to resort to doing this, I would buy a few litres at a time, enough to get me to the next town where I hoped I could find something more dependable. Each time I enquired, I was assured my search for clean fuel would be rewarded soon, but they were never sure when. The same thing usually happened when I asked someone in Africa how long it would take me to get to a place I was heading towards. The local people's knowledge of nearby landmarks seemed limited to about a fifty-kilometre radius. Beyond that, they rarely had any idea, as most had never travelled that far, and estimates for how long a particular journey would take were often wide and varied. It added a little spice to planning a destination for the day, which caught me out several times, finishing a tough, lonely ride well after dark.

Most of the primary roads were reasonably well-surfaced, but the sometimes more direct secondary roads were often in very poor condition. As I rode the last sixty kilometres of the journey from Mekele towards the small town of Lalibela, the road I had chosen got steadily worse, dipping into steep-sided gullies, over dry river beds, and through mud-hut villages full of excited children.

I began to wonder if my route choice had been a smart one.

Just as the sun dipped lower towards the horizon, a light rain started to fall. I had no idea whether the distances locals had quoted were accurate, but with the bike easily handling all of the terrain, I felt quite adventurous. I suspected that riding into the unknown was the norm from that point onwards.

While my maps were average, my compass was true, and I knew I simply needed to keep moving south.

CHAPTER TWENTY-FOUR

What Was Behind Me

Personal Diary Entry: 23rd March 2014

Day 314 (of 549)

44,069 km of 62,840 km (27,383 mi of 39,047 mi)

If the path to success was simply a matter of combining hard work with ingenuity, then Africa would be full of successful people – especially the women. I am frequently reminded of how I have squandered opportunities afforded me by where I was born and the colour of my skin.

The people of Ethiopia are beautiful. They proudly told me it was because they were the only African nation never to have been colonised. The most stunning women often propositioned me, but they were always refreshingly honest about their motives: a new life overseas and an opportunity to prosper. I tried to dissuade them, telling them my wife had just run off with another man because I was such a terrible husband – but that didn't seem to deter them.

The country challenged many of my preconceptions.

There was heartbreaking poverty everywhere, but also a cheerful positivity that belied the destitution and a can-do attitude that humbled me. It was anything but the war-torn, famine-ravaged wasteland I had seen through the lenses of our Western media.

Agriculture was the main employer, and most people still lived in the fertile rural regions. Farmer's markets were just that – actual farmers selling their crops. It was not uncommon to see caravans of camels and donkeys driven towards the nearest town carrying their seasonal harvest, sometimes a couple of days before the Saturday market. When the markets opened, there was a raucous festival atmosphere, fuelled by locally brewed alcohol and spicy snacks, as thousands gathered to trade and barter for all kinds of produce and goods.

Although the quiet mountain town of Lalibela held much more of interest than just its churches, it was these iconic structures for which it had become famous. Hewn from the solid bedrock and frozen in time, they were a sight to behold. Connected by an intricate network of pitch-black, subterranean tunnels and winding staircases, each one was still in use. Pilgrims descended on this town from far and wide to worship in the cool air within. Each day, around noon, a haunting chant echoed through the air as services commenced, and the faithful were called to gather and offer their prayers and devotions.

The surrounding mountains had endless opportunities for some great hikes, although by late afternoon, heavy clouds would build, and a huge electrical storm would unleash its raw power over the town. Power cuts were common and the water supply tenuous, but the locals took it all in their stride and candles were never far from hand.

I stripped my bike of its luggage and stored the heavy cases inside my room at a quiet guest house, which allowed me to explore some of the rougher off-road trails in the area. With the rear seat clear, I'd often give rides to locals walking between the small villages in the region. Sadly, I didn't have the space to carry passengers when the bike was fully loaded, so it was nice to have a little company at times.

The people I picked up seemed to appreciate it.

As I got to know a few of the Lalibela residents, I joked about my imminent "Ethiopian" birthday. These kind locals threw me a surprise party and celebrated with me. After attending a coffee ceremony in the morning, they invited me to return to the same modest home later that day. I spent the day hiking, curious about what I would find upon my return.

When I arrived, a freshly baked loaf of bread held several candles, and they gave me a homemade card and a hand-carved cross. Again, I was surprised that those with the least were often the most generous and deeply touched by this gesture, which reminded me to focus on being less selfish.

As always, the magnitude of my journey made me feel like I needed to keep moving.

After a wonderful week with incredible people, I loaded up my bike, said goodbye, and took the road to Bahir Dar on the shores of Lake Tana. Once again, I rolled through the spectacular countryside on delightful, twisting mountain roads before gradually dropping down into the flatter terrain surrounding Lake Tana, the source of the Blue Nile River. I stayed in the university town of Bahir Dar, considered by many Ethiopians to be their Riviera, with broad streets lined with palm trees and views of the sparkling waters of the vast lake.

On my first day in town, I took a walk out to the bridge overlooking the Blue Nile, where it exited the lake. Soon, I was snapping pictures of the abundant wildlife.

Downstream, a pod of hippos wallowed in the muddy waters while dozens of pelicans stood watch on the nearby shore. I couldn't help but notice several locals crossing the bridge and giving me strange looks, some shaking their heads and others wagging their fingers. When a policeman approached, I wondered if the Ethiopian authorities shared the same paranoia over picture-taking I had encountered in Sudan.

The police officer enquired as to what I was doing on the bridge.

With a sheepish grin, I pointed to the river and then to my camera, but as we glanced over the railing together, we noticed a group of naked men scrubbing themselves vigorously in the waters directly below. He gave me a suspicious glare and ordered me to move off the bridge immediately. *Whoops.*

Thirty kilometres from Lake Tana, the Blue Nile plummets over a waterfall known to locals as *Tis Abay*, which means "great smoke." I powered the motorcycle along a heavily rutted road, riding through long stretches of deep mud from the recent rains to the beginning of a trail into the falls. While it was a strenuous ride, wallowing in the wet dirt with my tyres caked in the thick, moist clay, the bike handled it all without too much trouble. After a short hike, I was soon standing at the base of the falls enjoying its cooling mist. A hydroelectric plant had diverted a considerable amount of the original flow, but it was still an impressive sight.

Lake Tana was also home to many island monasteries. However, with shifty boat operators hustling hard for business, I decided to explore the area on a mountain bike instead. I rented one from a hotel in town and took off around the shoreline. With no gears or suspension, it was a bumpy ride, and I would have been thankful for my own bicycle, which was gathering dust back in California.

Within a few kilometres of town, narrow tracks took me through the thick forest into villages where it felt like they had never seen a white person before. Children ran alongside the bike, laughing and waving, easily keeping up with my slow pace.

After four days in Bahir Dar, I packed up and took the road southeast to the capital, Addis Ababa. Yet again, I underestimated the distances, terrain, and conditions, and soon, I was racing against the clock, determined not to get caught out after dark, especially since finding dependable fuel was becoming more problematic.

I noticed few privately owned vehicles on the roads. Cities were swarmed by buzzing *tuk-tuks,* three-wheeled motorcycle taxis that ferried people around at minimal cost. The tuk-tuks competed for business with the larger mini-vans, which raced around, impossibly overloaded, inside and out. These had become a baneful nuisance as they often battled with each other for the next customer, stopping frequently without warning, usually pulling out with no regard for what was approaching.

Then, there were the white Toyota Land Cruisers, belonging to the plethora of charitable NGOs that thrived on Ethiopia's misfortunes. Mostly empty, they sped recklessly between towns during the day, only to gather outside the ritziest hotels by nightfall. I never actually figured out what practical purpose they served. To top it all off, copious amounts of diesel fumes and vast quantities of dirt spewed from large buses and trucks lumbering slowly along. Getting stuck behind one of these ancient behemoths left me sucking in lungfuls of smoke and sand before summoning the courage to blindly break out of the dust cloud and into oncoming traffic, hoping my timing wouldn't prove fatal.

I had one of the closest calls of the trip on the road to Addis Ababa as this very manoeuvre left me face-to-face with a stubborn bull who refused to move from the road. I missed one of his horns by inches, and he was a big enough beast that he would have probably ruined my day – not just my underwear.

Finally, as I approached Addis Ababa, the dark clouds building all afternoon released their contents of thunder, lightning, rain, and even snow – yes, *snow* – that accompanied me on the final descent into the capital. By the time I reached the city, darkness had fallen, and the deluge had overwhelmed the streets. I was left cold, sodden, and hungry as I attempted to navigate the way to my lodging. Rivers of refuse raced to the lowest point in the city, followed by the foetid odour of raw sewage.

My first impression of Addis Ababa was not going so well.

Additionally, workers had torn up large portions of the city centre to accommodate a Chinese-sponsored railway project. When I stopped and asked for directions, even the locals were confused. Exhausted, I finally found my destination – a cheap backpackers hostel on the south side of the city. As I checked in, large puddles formed around my feet on the reception floor, the rain oozing out of the creases of my motorcycle gear.

Many have often asked why I undertook this journey, and on such days, I sometimes asked myself that same question.

After a day on the south side, I moved closer to the city centre and pitched my tent in the lush gardens of Holland House, an overlander's haven tucked discreetly behind an old bus station. As I pulled my motorcycle into the secure compound, I noticed another bike, almost identical to mine, parked under a fruiting banana tree. It belonged to a young English rider named Ross, whom I had heard stories about from the BMW crew in Egypt. He was on his own adventure from London to Cape Town, so we had plenty to talk about over a beer later that evening.

I hadn't been there more than an hour before another two F 800 GS motorcycles turned up, coming from Cape Town with a friendly German couple, finishing the last stages of their "round-the-world adventure." Suddenly, we had four of the same motorcycles parked at Holland House and plenty of stories of how our bikes and bodies were holding up over our respective journeys.

Our first night together lasted well into the early hours of the following morning.

Over the next few days, we all completed some overdue maintenance while comparing notes on any small issues we had been having with the bikes. Overall, it was a hearty thumbs-up for the F 800 GS. We each had concerns over the new, untested machine, as it had not been on the market long enough to gain the reputation of some of the older, proven overlander bikes. Still, all of us were impressed by how tough and capable they were in all kinds of terrain.

St. Patrick's Day just happened to coincide with our little gathering in Addis Ababa, and Wim, the Dutch owner of Holland House, took it upon himself to guide us on an exhaustive tour of the city's liquor stores in search of a bottle of Irish whiskey. Not a single drop was to be found, other than behind the bar of the plush Sheraton Hotel, so we settled for some local beer instead.

I had been in Ethiopia for several weeks by then, and I had thoroughly enjoyed every second of it. The locals made my stay a memorable one, but I felt I still had much more to see. The roads were exceptional, although who knew how durable they really would be? Trucks overloaded well beyond their intended capacity were already causing noticeable damage. But off the main highways, there were still plenty of dirt road adventures to be had. The terrain was breathtaking and always accompanied by a variety of flora and fauna.

I had never imagined Ethiopia would be so interesting and diverse.

Cheap and cheerful, it was the perfect location for the more adventurous traveller, but a question mark still hung over how well they would manage the tourism industry. Sustainability seemed to be a low priority at many of the popular destinations, and the erosion from overuse was beginning to show. With a motorcycle, the entire country was open for exploration, and getting off the beaten track was as simple as turning the handlebars.

As I planned to move further south, I knew I would miss what was behind me, but if I slowed my pace any further, I doubted I would ever reach Cape Town.

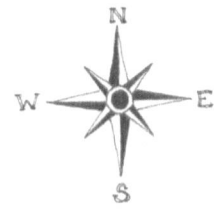

CHAPTER TWENTY-FIVE

Throw Caution to the Wind

Personal Diary Entry: 18th April 2014
Day 340 (of 549)
46,685 km of 62,840 km (29,009 mi of 39,047 mi)

As the storm of emotions that still rages within me slowly begins to subside, I catch glimpses of a new horizon emerging. It promises the potential for growth, but my trajectory is still not clear. I struggle to fully surrender, fearing what will happen if I let go. I'm not sure if I possess the strength and resilience to navigate this tumultuous juncture, and yet I suspect that if I leap, the net will appear.

There were two possible roads from Ethiopia into Kenya. The most commonly used route went through the border town of Moyale. With tribal conflicts in the area creating occasional problems for those passing through and frequent kidnappings and gunfights, everyone recommended I only travel during the daytime and, ideally, as part of a large, armed convoy.

Some time ago in Egypt, another rider had suggested I attempt the alternative route from Omorate to Lake Turkana, but he had warned of the remoteness of the region with potentially no fuel, food, or water for up to 1,000 kilometres and laborious riding conditions along the entire way. When riding conservatively, my bike's range was approximately

650 kilometres, so I'd decided that the Turkana route would be an unwise choice if I couldn't find someone going in the same direction.

As it turned out, my new acquaintance from Holland House in Addis Ababa, Ross, had similar plans. He was also waiting for someone to ride along with, so we sat down, shared a cold beer, and discussed how best to approach this more complicated route. The lack of fuel would be our most pressing concern, followed by water and food. With my bike already carrying close to its maximum load, I was concerned about adding more weight. The rear shock on the F 800 GS was said to be one of the bike's weaker components and Ross had just replaced his after it had failed several weeks earlier.

Ideally, we wanted to find a group of four-wheelers going in the same direction so we could offload the additional fuel and water needed to complete the journey. Overlanders passed through Holland House with regularity on the long road from Cairo to Cape Town.

Within a few days of us hatching our plan, a large Toyota Land Cruiser pulled up, carrying two young men on their journey from the Netherlands to South Africa. We did our best to convince them to join us on our intended route, but their tight schedule would not allow for the significant detour. The momentum of our planning and the lure of the unknown swayed our preference towards the more interesting Turkana route – even with its logistical difficulties.

We decided to recklessly throw caution to the wind.

With our bikes packed, we said goodbye to the wonderful staff at Holland House and turned south along the Rift Valley, swapping the grimy city streets of Addis Ababa for the pristine lakes and mountains of Southern Ethiopia. We stopped briefly in the Rastafarian outpost of Shashemene before pulling into the lakeside town of Hawassa in time for a cold beer with some friendly Peace Corps volunteers as the sun set over the placid waters. Impossibly large marabou storks quarrelled in the trees above us and scavenged along the shoreline while the sky blazed with colour.

The following morning, we left the peaceful town and headed southwest towards the small city of Arba Minch on the shores of Lake

Abaya, the last sizable settlement before we would leave the paved roads and venture into the Omo Valley region. We spent a couple of days there stocking up on supplies and filling every container with additional petrol, wondering where we would find the next source of fuel.

Soon after leaving Arba Minch, the asphalt disappeared, heralding the beginning of the infamous thousand kilometres of dirt roads. The surfaces varied dramatically from bone-jarring corrugations to well-graded, smooth dirt. We had both fitted new rear tyres before leaving Addis Ababa, and we were glad of the additional traction provided by the aggressive tread patterns.

Progress was fast, so we took a short detour.

We couldn't help stopping at the small village of Key Afer, where a colourful open-air market was in full swing. Tribes from the local region gathered in their traditional dress adorned with colourful beads and animal hides, their hair braided and coated in red ochre and animal fat. As soon as we pulled up on our large motorcycles, we became the centre of attention, and a crowd of curious onlookers quickly surrounded us. Children poked and squeezed our bulky riding outfits, marvelling at the underlying armour that made our elbows, knees, and shoulders look somewhat deformed. It was oppressively hot underneath our heavy suits, so we stopped for a cold drink at a nearby guesthouse before resuming our ride towards the village of Turmi and the camp we had chosen for the night.

Just before reaching Turmi, on a stretch of particularly rough road, the back of my motorcycle began to violently fishtail. I assumed I had blown out my new rear tyre. Like a bull rider, I held onto the bike so it wouldn't throw me off while I attempted to keep it upright. When I managed to bring it to a halt, I found that one of my aluminium panniers, a storage container that comes in a set attached to both sides of a bike, had detached itself from the luggage rack and was dragging behind the bike, held on only by the thin security cable. As a result, one of my spare fuel containers had ruptured. After salvaging the remaining contents, I carried out a quick roadside fix.

I noticed a man on the side of the road, staring, awestruck.

Every time we pulled over along this road, no matter how isolated we thought we were, a local mysteriously turned up to gawk at our unusual machines. Sometimes, they were small children – too young, I felt, to be standing unsupervised on the side of the desolate road, but this was a different world than my own. Often, these locals came up to us and attempted to interact with us somehow, even though they rarely knew any English words.

That man did not come near – he just watched safely from a distance until we were on our way again. Soon, we reached a small campsite on the outskirts of Turmi that a Dutch couple had recommended. They had travelled through the region a few weeks earlier. It was getting late, and no locals inexplicably appeared like apparitions.

We were the only ones there.

After a peaceful night, we woke and began to pack up our camp. Then, we noticed a young boy from the local Hamar tribe deftly climbing through the mango trees above us. When he reached a comfortable limb to sit on, he began showering us with ripe fruit for a delicious breakfast.

From Turmi, we continued southwest towards the sleepy border town of Omorate to have our documents stamped out of Ethiopia. On our arrival in Omorate, as we negotiated with some locals over the price of the black-market fuel, the skies started to darken. Suddenly, a huge rainstorm began pelting us with fat drops of moisture. Almost instantly, the sandy roads turned into slick, sloppy rivers of mud. The locals warned us that the roads in town were nothing compared to what lay ahead. They recommended we spend the night so that the tracks could dry out before we got into the really difficult terrain. The rainy season was upon us, and as we settled into a cheap, simple hotel, we wondered if attempting this route was a wise choice at this time of year.

By morning, the roads appeared to have dried.

We decided to continue, first pausing to fill our fuel and water containers. Contrary to what some had told us, it seemed as though fuel was not as scarce as we'd thought. Soon, we set out to find the indistinct turn-off towards Kenya.

Immediately greeted by deep sand, muddy river crossings, and blistering heat, it was slow going, and by early afternoon, after nearly six hours of strenuous riding, we had covered just over seventy kilometres. Exhausted, thirsty, and hungry, we pulled into the tiny outpost of Ileret, just inside the Kenyan border. The prospect of several more days of this type of travelling was daunting and exhilarating. We registered with the local police, the accepted procedure in the absence of an official border post, before locating a small shop where we enjoyed a warm Coca-Cola and a packet of dry biscuits.

The police kindly offered to let us use their barracks as a camp for the evening. Then a local man from the Daasanach tribe suggested we stay at the Turkana Basin Institute (TBI), ten kilometres to the south. We took his advice and rode to an unattended gate, hoping to see someone in the compound behind it. The surrounding chain-link fence disappeared into the bleak, interminable landscape, and a rough airstrip outlined by rusting fuel cans paralleled the faint road we had ridden. The parched earth cracked beneath the unforgiving sun. There was no one in sight.

"Do you think this is the right spot?" I asked Ross.

He shrugged his shoulders. "I can't see what else it could be."

We rattled the gates and shouted into wind, hoping to attract somebody's attention. Before too long, a faint silhouette materialised on the horizon, deep within the compound. Someone was approaching. The ethereal figure slowly solidified into a handsome young man wearing a clean white shirt, formal trousers, and sandals made from a car tyre.

"Hello, my name is Joseph. How can I help you?" he asked politely when he finally reached the gate.

We introduced ourselves and told him about our encounter in Ileret with the man who had directed us to that location. We asked if it would be possible for us to stay inside the compound.

"Let me speak to Mr. Richard. He is the boss. He is in Nairobi. I must call him on the telephone."

Then, Joseph turned and casually walked back in the direction he had come from, his form slowly disappearing into a shimmering speck hovering above the endless sand.

Ross and I looked at each other and smiled.

We took off our jackets and settled into the shadows created by our bikes. The sun slipped lower in the sky, and our shadows grew. As soon as we began to doubt whether we would ever see Joseph again, his lonely figure reappeared. We stood to greet him and hear his news.

"Mr. Richard said you are most welcome at the institute. Please, come." Joseph took a key from his pocket and unlocked the gate, allowing us to enter.

Joseph was the manager, he explained, as gave us a tour of the extensive complex of partially finished buildings. It was only after asking him the purpose of the institute that we realised we had stumbled across one of the dig sites of the renowned Kenyan paleoanthropologist, Richard Leakey. Still under construction, the site would eventually host fieldwork groups from around the world as they explored the fossil-rich region more extensively. What they had found to date strongly suggested this region was pivotal in the evolutionary story of humanity, commonly referred to as the "Cradle of Mankind."

The staff at the institute made us feel truly welcome.

We dined with the workers that evening on a simple, delicious meal of lentils and chapatis and camped under a sky so bright with stars it didn't seem real. The following morning, we awoke to the comforting aroma of sweet, milky spiced chai. Over breakfast, we thanked the crew for their gracious hospitality before mounting our loaded bikes and waving farewell. Our machines roared to life, and we exited the compound, determined to maintain our momentum. With hearts full of appreciation and a sense of adventure dancing in our souls, we rejoined the infamous Turkana Road.

But fate had other plans for us that day.

Less than ten kilometres from the gate, Ross dropped his bike after a patch of deep sand caught his front wheel and threw him off balance. As he wrestled it from the ground, I stopped, shut off my engine, and ran to help him lift it. Our bikes were so heavily laden that it required a Herculean effort to get them upright after a spill. When I tried to restart my bike after he was back in the saddle, it wouldn't respond. It was like the battery had suddenly lost all of its charge, and my heart sank further with each failed attempt to start it.

The ever-resourceful Ross rode back to the institute and asked for some assistance. Soon, an ageing Land Cruiser appeared with Ross leading the way. We threaded a rope through its engine protection bars and around the truck's tow hitch to pull the bike and keep it balanced. As the driver accelerated, towing my bike and me back to the compound at high speed, it took every ounce of my strength to stay upright as my front wheel ploughed through the deep sand. I wanted to wave at the driver to get his attention, hoping he might slow down, but I couldn't release my death grip on the handlebars.

It was an exhausting and terrifying ten kilometres.

It was still early in the day, and we spent the rest of the morning going through probable causes for my sudden loss of electrical power. When I say "we," it was mostly Ross who diagnosed the bike. As a professional engineer, he had an intimate knowledge of all things mechanical and electrical. By lunchtime, he'd isolated the fault to a failed stator within the alternator.

The alternator was an essential component, generating electrical power for the bike. While the engine was running, it would top up the battery's power, which, in turn, served the bike's multiple electrical systems. The stator, nestled snugly inside the alternator, resembled a doughnut wrapped in fine copper filaments, and mine was clearly overcooked. Without power, the bike's Engine Control Unit (ECU) ceased to function, the fuel injectors quit, and the coils failed to create their much-needed ignition sparks. Unfortunately, we were a long way from any kind of repair or replacement opportunities.

We were stranded.

Joseph made another call to the head office of TBI, and they kindly agreed to allow us to stay for as long as it would take to have a spare part brought in or to have the bike carried out on the back of a truck.

Meanwhile, they authorised the use of their well-equipped workshop. Over the next few days, we looked for a viable solution or a suitable truck that could fit the bike on board for the long journey out of the region. We were at least a three-day drive from Nairobi, and at the end of a long and gruelling road.

Trucks were rare in this region, and all spare seats were typically oversubscribed long before they even arrived. There were rumours that a large truck was making its way to the institute from the town of Marsabit, but it was still two days away, and road conditions were preventing it from getting closer. When a truck finally did show up *four* days later, the driver wanted an extortionate amount of money to take the bike to a place where it could be repaired or stored until they located a spare part.

In the meantime, I had purchased a used 12-volt truck battery from Father Michael, a German priest who served at a remote Catholic mission deep in the heart of this inhospitable region. I disconnected all non-essential electrical functions on the bike and used thick cables to hook up the respective battery terminals. I then fastened the cumbersome load to the back seat of the bike – and it worked!

Happily, we set off once again.

We had spent several nights at the institute, and they had looked after us well. It was hard to say goodbye, but we were determined to get out of the area under our own steam. For the next five days, we struggled south on some of the most formidable roads I had ridden to date. Deep sand – my personal nemesis – loose volcanic rock, slick mud, and thorny bushes all compounded to make for a true adventure. We were never sure how long the battery on my bike would last, and we spent many hours on the side of the road, switching power sources from one bike to the other.

To our surprise, the virtually non-existent road was not quite as desolate as we had been led to believe. In the evenings, we pulled into small villages, hoping to find a charger to top up the large battery. Unfortunately, virtually all of the chargers were solar-powered and of no use to us at night, so we spent extra daylight hours waiting for adequate power.

Our concerns over the lack of fuel took a backseat after we discovered the few villages we travelled through had at least one, albeit overpriced, source. What was more worrisome was how the bikes were taking a beating. After over 1,000 kilometres of rough roads, the damage was considerable: one fried alternator, two flat tyres, one battered pannier, and a cracked subframe. Each time something broke, we found a way to fix it, improvising a repair with only what we carried.

In hindsight, it was all worth it.

The terrain we experienced along the way was breathtaking – lush, rolling hillsides, scorching hot, barren deserts, deep, jagged gorges, and the basalt rocks from the extinct volcano, Mount Kulal, strewn along the shores of Lake Turkana. Wild herds of oryx and zebra roamed the open plains, camel trains crowded the dusty roads, and the people we met were simply fascinating. Many of the experiences we had would not have happened without our challenges along the way.

When we hit our first asphalt road in Kenya, we were glad to finally ride on a smooth surface, though a little disappointed that our only significant danger from now on would be other drivers. However, I was more of a danger to others, being a non-local, and needed to keep reminding myself that people drive on the left in Kenya, unlike in Ethiopia.

We broke up the ride to Nairobi by stopping at a campsite near the town of Nyahururu, where we bumped into an inspiring young Dutch couple on their voyage to Cape Town. Ross had ridden with them in Libya, and they were coincidentally staying at the same campground. Such an auspicious reunion gave us the excuse to celebrate with our first genuinely cold beers since leaving Ethiopia.

The next day, after charging my battery for the final time, all four of us set out along the edge of the Rift Valley, across the equator, and into the bustling, sprawling city of Nairobi. We planned to drive until we reached the comfortable overlander haven of Jungle Junction.

I knew I would have to spend a few days there while I found a local expert with experience in repairing burnt-out stators, which I hoped would be significantly cheaper than having a new one shipped from abroad. My bike was long overdue for some thorough maintenance, and with two full-time mechanics onsite and the knowledge of the resident expert Chris Handschuh, I couldn't think of a place more suited to my needs than that particular campsite.

Just as I pulled into the premises, my bike spluttered to a halt. The battery was completely spent.

CHAPTER TWENTY-SIX

An Oasis of Kindred Spirits

Personal Diary Entry: 4th May 2014

Day 356 (of 549)

47,292 km of 62,840 km (29,386 mi of 39,047 mi)

It has taken time and an ongoing conscious effort to let go of my attachment to certain outcomes, to realise that many of the tributaries of my life are beyond my influence.

I'm coming to terms with a deeper wisdom: the joy of living lies not in orchestrating all the details but in letting the unpredictable currents shape my path.

With the Easter holidays coinciding with my arrival at Jungle Junction, it became clear that dropping off my damaged alternator with a local mechanic would prove pointless. It would probably just sit on a workbench, gathering dust until after the long weekend, before eventually receiving any attention.

I pitched my tent and settled in for a long wait.

If I could have chosen one place to get delayed in Kenya, then Jungle Junction, on the outskirts of Nairobi, would have been my first choice. Long established as an overlander's sanctuary on the route through eastern Africa, it became an oasis of kindred spirits, beaten and battered by the tough roads to the north and south. Amid the scent of oil and camaraderie, a tapestry of stories unfolded; strangers became confidants, united by wanderlust.

With well-trained mechanics always on hand to offer advice and assistance, a fully equipped workshop, and the friendly owners, Chris and Diana, opening their home to the likes of me, it was a welcome respite from the chaotic world outside its well-guarded perimeter. Travellers from around the world gathered here to regroup, repair, and recover. Rows of overland vehicles quietly sat awaiting the resumption of their adventures, temporarily stored on the expansive property while their owners were overseas rebuilding finances or accruing the time off needed to continue their journeys.

The region was now well into its wet season, which wasn't the easiest time to travel. But, because of this, the campground had attracted most of us who had decided to travel during this quieter period. Many roads were considerably more arduous or simply impassable, and Jungle Junction, located roughly halfway between Cairo and Cape Town, proved to be an ideal rest stop. During my stay, I met more people than I had expected.

These people got me.

They understood the motivation behind jumping on a motorcycle and driving so far. No mishap or obstacle was insurmountable for us. Breaking down just meant a longer stay, running out of funds meant that we'd be back in a little while, and loneliness meant we just needed to make a new friend to ride with. Ending up suddenly single meant I had absolute freedom to choose whatever path appeared before me.

Most conversations about my journey with people outside of this community normally ended with puzzled looks and consternation. I had been asked more than once why I didn't just fly to Cape Town.

Our species has been more restless than not.

It has only been since the agricultural revolution, roughly 12,000 years ago, little more than a blip in our evolutionary timeline, that humankind has favoured settlement over a more migratory existence. I believe we are travellers from birth and that movement is the very essence of life. I recognise the urge to keep moving that is hard-wired into my primitive brain.

Over time, we have created an economic system that thrives on dissatisfaction and misery. I've often wondered if the obsession for comforting ourselves with endless amounts of "stuff" that we don't really need and rarely use is an effort to compensate for the restraints we have willingly accepted. Are the stresses and maladies of our frantic lifestyles a consequence of a much deeper problem? If we were truly content, would we really need that next "thing" that promises to bring us happiness? Would we trade our limited time for something ultimately unattainable – a sense of contentment through acquiring possessions?

It was hard to explain why it felt so right to be on this voyage.

Looking around at the fellow travellers who gathered at Jungle Junction, I could tell they felt the same. There appeared to be a common sense of purpose and drive among them. They understood the meaning of life. They, too, had the desire to leave most of their possessions behind and travel frugally through barren deserts, hoping not to run out of supplies but trusting that, if they did, someone would show up, willing to help. Like me, they didn't mind cold showers or no showers at all for a while, sleeping on the hard ground, or the discomfort of riding through a rain or dust storm.

Such experiences filled us with stories. The truest joys in life are not found in material things. Love, laughter, and shared experiences are some of the simplest treasures that enrich our existence. As a result, camaraderie came easily, and we formed friendships instantly. There was a thirst for adventure and a palpable vitality surrounding everyone. Everything just felt right, no matter the setbacks each individual might have experienced before they got to this gathering of like-minded people.

***

The Easter weekend passed quietly. A barbecue on Good Friday evening saw a dozen overlanders enjoying great food and cold beer. Days were consumed with routine maintenance procedures on bikes and trucks, the packing and unpacking of gear, the consolidation of equipment, and the sharing of advice about the roads ahead. The hot, humid days were followed by cool nights and the occasional thunderstorm. After the intense heat of the northern regions, it was nice to sleep peacefully without constantly sweating.

By Tuesday morning, businesses in the city were coming back to life. I took my alternator to an electrical engineer at the regional airport to have the doughnut-like stator rewound with thin copper wire. A local biker had recommended this particular engineer, and I assumed that anyone who worked on aircraft would complete the task to an acceptable standard.

From Jungle Junction, I took a *matatu,* a minivan licensed to carry twelve passengers, towards Wilson Airport, near the city centre. Matatus were the most common vehicles on the road – and the least predictable. They could be flagged down at any time, and the driver's income was directly related to the number of people on board, so they were motivated to drive fast and stop in a heartbeat to squeeze in another person.

The matatu I took had a few extra people onboard – I lost count at twenty-four – but for mere pennies, these ramshackle vehicles would take passengers across the city on a rollercoaster ride of emergency stops, screeching brakes, and blaring music. Road rules didn't seem to apply to the matatus, so anything was deemed acceptable in the race to get that next passenger. It made for an exciting, if somewhat fragrant, experience.

The engineer's workshop at the airport looked clean and efficient, so I felt confident about leaving my stator in his care. A timescale and a price were to be agreed upon later after he'd had a chance to thoroughly inspect the damage, so I readied myself for the return journey to Jungle Junction.

Many overlanders I met previously had left me with the impression that driving through Nairobi had been one of the most dangerous parts of their journey, and I was tempted to concur. Unfortunately, it was virtually impossible to drive around the city, so all through traffic was channelled into its centre. The government was building a ring road, but it would be several years before that project was complete, so, for the time being, the city's centre was in a permanent state of gridlock. Traffic lights were ignored, roundabouts were a free-for-all, and motorcycles were at the bottom of the food chain.

The city also had a rough reputation for crime, and I'd heard it referred to as both "Nowrobme" and "Nairobbery." Muggings were common, even in the daytime, and by nightfall, the city centre was a no-go area for tourists. In stark contrast to the modern, glass-tower blocks, the city was peppered with shantytowns, the largest of which was Kibera. Thought to be the biggest urban slum in Africa and the third largest in the world, estimates varied as to how many people lived in the squalid township.

The most common complaints I'd encountered from the locals in Kenya were directed towards the scourge of corruption, endemic within the world of officialdom; it infected every aspect of life within the country. Very little happened without a bribe of some sort greasing the palm of someone in authority.

When it looked like I would have had to transport my bike out of the Turkana region, the driver I was negotiating a price with had demanded almost 1,000 pounds to get me and the motorcycle to Nairobi, which amounted to roughly one-third of the average annual income in Kenya. Not five minutes later, he was grumbling about greed and corruption within the government, and I had to bite my tongue to prevent myself from pointing out the hypocrisy of his complaints. Luckily, I didn't need to use his "reasonably priced" services. It took four days to repair my stator, and after fitting it back inside the engine casing, I ran a few tests to check its output. It appeared as though it was working well.

I knew only time would tell.

Having a better means of getting around than a matatu, I had some errands to run. Until then, I'd been doing my best to avoid any contact with those in uniform. But, since I had entered the country at a border with no official immigration or customs outpost, I needed to visit the immigration headquarters in downtown Nairobi to be legally present in Kenya. When I arrived, I expected them to ask for a bribe to facilitate the bureaucratic formalities.

Surprisingly, it all went smoothly.

With a working bike and my paperwork in order, it was time to make plans to move on before I got too comfortable at Jungle Junction. The Handschuhs took good care of me and the other travellers who stopped in Nairobi to rest on their land. With Chris helping with repairs and filling me in on what to expect in the area, Diana cooking breakfast that wafted through the campground, and refrigerated beer often handed to me by someone or another, I was a little reluctant to leave. But the travelling companions I had arrived with had long since departed, and after almost two weeks, I was also keen to get back on the road. As I rolled up my tent, the pale grass I exposed served as a reminder as to how long I'd camped in Nairobi.

After consulting my maps and taking some advice from fellow riders, I decided to turn west and enter Uganda via the northern edge of Mount Elgon. I'd completed several major repairs to the bike, so it had been productive. It had also been an enjoyable stay – but it was time to go.

***

Remember when I said I'd been doing my best to avoid any contact with those in uniform? My strategy was especially true when it came to avoiding run-ins with local police. There was a heavy police and army presence within the capital to counteract the occasional grenade attacks for which Al-Shabaab militants had been blamed. As a result, I often encountered random police checkpoints on the road.

Well, I had a method for getting away from the police.

When they saw me approach, they quickly became animated, jumping into action. As they waved me over onto the hard shoulder to shake me down for money, I simply waved back and kept on driving. They rarely had a car nearby, and when they did, they seemed reluctant to use it. Luckily for me, they never followed me.

After a tense drive out of the city into the surrounding hills, I was soon competing for road space with suicidal matatu drivers. As the road narrowed from four lanes to two on the way to Lake Bogoria, a flamingo sanctuary within the Rift Valley about 200 kilometres northwest of Nairobi, I was forced onto the hard shoulder several times to avoid a head-on collision with an oncoming vehicle. Thankfully, my chosen route eventually took me off the main highway. I was soon speeding along deserted smaller roads, waving at the occasional police officer as he tried to flag me down.

Playing dumb kept me out of trouble.

I arrived at a campsite by late afternoon, and after checking in, I attempted to restart the motorcycle, but it died immediately. The battery was strong, so I knew it couldn't be the newly repaired alternator. With night approaching, I decided to deal with it after a good night's sleep. The following morning, I traced the problem down to a fuel starvation issue. When I bypassed the fuel pump regulator and wired the battery directly to the fuel pump, I was able to get the bike running again. However, that wasn't going to work long-term.

I abandoned my plans to continue west. I turned back towards the glorious anarchy of Nairobi so I could utilise Chris Handschuh's intimate knowledge of this bike and all its little intricacies. Before establishing Jungle Junction, the overlander's haven, Chris had worked as a BMW motorcycle technician in his home country of Germany.

I was disheartened to have the machine I had relied on for almost 50,000 kilometres suddenly develop mechanical issues. The majority of riders who attempted this route went from Cape Town to Europe, so, as their mileage accumulated, their journeys took them closer to more developed countries with readily accessible spare parts and expertise. The faith I had developed in the bike's reliability was slowly being chipped away, and I wondered about some of the more remote route choices I was considering on my journey south.

I was learning to accept the setbacks as opportunities to understand more about the bike and myself. It was frustrating, but I reminded myself that it was only a machine and I was only human. Things would occasionally go wrong, but it wasn't the end of the world.

An expression I had heard more than once since arriving in Kenya was the Swahili phrase *hakuna matata,* which roughly translates to "no worries."

CHAPTER TWENTY-SEVEN

If It's Not Broken, Don't Fix It

Personal Diary Entry: 20th May 2014

Day 372 (of 549)

48,441 km of 62,840 km (30,100 mi of 39,047 mi)

Measuring a journey in terms of distance covered or time elapsed provides only a superficial gauge as to what a voyage is truly about; it is the unique combination of experiences along the way, the highs and lows, the triumphs and disappointments that really give it life.

When I reunited with Chris in Nairobi, the motorcycle guru gave me some Teutonic words of wisdom that reminded me a lot of the advice Zeb offered Sarah and me at the beginning of the trip: "If it's not broken, don't fix it."

Me being me, I spent a couple of days switching parts around, testing, and retesting, before reassembling the original components and trying it one more time. To my surprise, everything worked perfectly, but I was left with the frustrating dilemma of not knowing what had caused the original malfunction.

Nevertheless, I gave Chris a thumbs-up and nervously decided to continue onward.

Before leaving the city, I took one last fateful trip into its chaotic centre. A late lunch in the downtown district with an old friend turned into a long discussion about Kenya, politics, corruption, tourism, corruption, gender equality, corruption, religion, and corruption. Before I knew it, the sun had set, and I was in downtown Nairobi after dark, a city with a shocking reputation for daylight robberies.

As my friend and I walked back to my bike, I thought I looked brave by resisting the temptation to wear my motorcycle helmet and gloves. On a positive note, some beautiful women suddenly approached me and seemed genuinely interested in spending time in my company. Flattered by the sudden attention, I soon had my ego deflated by my friend, who pointed out that these ladies were neither interested in my rugged good looks nor my charming personality, only in how quickly we could exchange bodily fluids for hard cash.

Back on my bike, I was soon weaving my way through the late-night traffic.

After riding through Nairobi at various times, I had established that there was no such thing as a rush hour in this great city. Nothing could rush, ever. It was gridlock all the time through the downtown streets, but, at least on a motorcycle, it was possible to make some progress. As I neared the edge of the city and the pace of traffic began to pick up, unbeknown to me, I was driving towards one of the milestones I had hoped to avoid altogether: my first crash.

Traffic lights governed roundabouts in Nairobi, which defeated the purpose of having roundabouts in the first place. I could only assume there was some logic to the sequencing, but locals interpreted each colour as a signal to either go or go *faster*. Adding to the confusion, a police officer often stepped into the mix to "direct" traffic, but on this particular evening, the officer on duty seemed too busy dodging the cars that were ignoring his frantic hand gestures.

I consider myself to be a cautious driver.

I always try to predict the worst-case scenario when scanning the traffic in front of me so I can be ready to react. The particular manoeuvre that caused me to crash was so far outside the realms of logical behaviour that I didn't see it coming. As I entered the roundabout, the driver of the matatu I was following decided the wide open space in front of him would be the perfect place to disgorge his passengers, so he came to a sudden stop in the middle of the busy intersection.

I swerved to avoid hitting the rear end of the minivan, missing it by inches, but as I turned the bike to correct my new trajectory, my front wheel caught a piece of debris left behind from a previous accident. Losing traction under the front wheel while already at a precarious angle could only result in one outcome, and I was soon sliding on my side to an inglorious stop. Before the police officer could think of a reason to issue a ticket, I picked up the offending debris, threw it off the road, checked the bike for damage, and got back on.

Somehow, I didn't get run over.

As I completed my journey home and the adrenaline wore off, I began to flex each limb, checking for tenderness and pain. I had no significant damage other than a few bruised ribs, a torqued back, and several scratches. A daylight inspection of the bike revealed the same. All the protection bars I had fitted to protect the vital parts of the motorcycle performed perfectly.

With that recent reminder of how vulnerable I was on a motorcycle, especially in a city like Nairobi, I packed my gear and prepared to leave the densely populated capital for a second time. I took a similar route into the Great Rift Valley but chose to camp along the shores of Lake Baringo, a freshwater lake famous for its birdlife and hippopotamuses. As I set up my tent, several curious vervet monkeys came to watch while colourful birds of all shapes and sizes fluttered throughout the canopy above.

It was only after dark when the hippos made an appearance.

Hippos spend most of their days wallowing in the cool waters, only venturing onto land at night to graze along the shoreline. At around midnight, I was awoken by the deep grunts of a large hippo pod as it came onshore, loudly munching grasses alongside my tent. Often weighing over 1,500 kilograms, they need to consume roughly one-tenth of their body weight each night, so the ensuing feeding frenzy kept me awake until dawn.

On more than one occasion, I checked my tent for anything I thought might attract their attention, vividly aware that only a thin layer of nylon separated me from these immense and often temperamental beasts. By morning, they had retreated to the water, from where they quietly watched me as I packed up. I was surprised to discover how close their footprints had come to the edge of my tent.

I made an early start that day, climbing out of the west side of the Great Rift Valley towards the small town of Iten. Once home to one of Kenya's former presidents, the roads were in remarkably good condition, and I was soon lost in a rhythm of twists and turns, climbing and falling over spectacular ridge lines through deep valleys lined with coffee plantations.

Crossing into Uganda was relatively easy, even though a swarm of fixers surrounded the bike as soon as I pulled up, assuring me that the process could only be completed with their costly help. My stubborn disposition and my meagre budget forced me to refuse their assistance; I was determined to navigate the border formalities alone. Fortunately for me, one of the few positive legacies of British colonisation was that most signage was still in English, so I was able to locate each of the offices by myself. After a sweaty couple of hours, another frontier quickly disappeared behind me.

I noticed an immediate improvement in road conditions and driving etiquette, but dark clouds hung ominously overhead. Before long, the first few raindrops were splashing onto my filthy visor, mixing with the dust and grime and making it almost impossible to see the road ahead. I pulled over at a fuel station and joined a large group of local bikers sheltering beneath the huge awning.

My 800cc bike attracted a lot of attention.

It was a rare occurrence to encounter motorcycles with engines larger than 125cc. Soon, I was answering the standard list of questions about my bike and the journey. The maps I'd painted on the side of my luggage panniers provided a useful aid for explaining where I'd come from and where I was planning to go. Eventually, the rain eased and the thunder and lightning abated, so I decided to push on while I still had daylight. I verified some rough directions and soon realised that Ugandans really wanted to be helpful. Even if they had no idea where a place was, they would happily make something up.

The strategy I'd employed for route planning had evolved considerably since the beginning of the trip. The guidebooks I carried were often inaccurate or outdated, so I followed the suggestions of people I met along the way, typically those coming from the regions I was about to enter. That new approach served me well and took me to some hidden gems I would have otherwise ridden past. A rider I'd met in Nairobi had suggested that a detour up to Sipi Falls was well worth the effort, so shortly after crossing the border, I turned north towards the infamous Mt. Elgon. I remembered reading that the caves peppering the mountain's vast flanks were once thought to be the source of the dreaded Ebola virus.[21]

Cloaked in a mysterious blanket of mist, Mt. Elgon failed to reveal itself.

With the wet season relentlessly drenching its summit, waterfalls cascaded over every exposed cliff face. Thoroughly soaked, I pulled into a small eco-camp on the edge of Sipi village, and while the rain steadily poured, I enjoyed my first Ugandan beer. I spent the following day exploring the nearby falls. Caked in sweat and mud, I clambered over rocks to the base of the tallest one. High above the valley stood Sipi Falls, thundering torrents of water gushing from the densely vegetated mountainside. I stared in awe as its mist fell upon my face, quenching my thirst and cooling my body in the oppressive midday heat. I couldn't resist pursuing the creek further upstream through the humid rainforest and sporadic coffee plantations.

From the village of Sipi, I retraced my route back to the south before turning west towards the town of Jinja. The Nile River had been a constant companion since I'd entered Africa. From the lazy waters of lower Egypt to the confluence at Khartoum to the Blue Nile source in Ethiopia, it had been a helpful landmark and a willing guide. As I neared the town of Jinja, the impossibly vast Lake Victoria stretched off into the far distance. I smiled to myself as I realised I had finally arrived at yet another milestone – the source of the White Nile.

For decades, the White Nile has drawn whitewater enthusiasts from around the world. Offering a challenging selection of big-volume rapids, it had long been on my bucket list of rivers to paddle. Sadly, a series of planned hydroelectric dams would see this stretch of river soon become tamed, controlled, exploited, and forgotten, but it still had plenty to keep me entertained.

Where there was whitewater of this quality, there would be rafting.

And where there was rafting, there would be river guides. It didn't take long to track down some of the guides I'd worked with in the past, and soon I was at the local bar catching up with some old friends. As usual, that first night went on until the sun came up. I felt quite exhausted after the reunions and revelries, so a relaxed afternoon cruise on the upper river helped to ease in the following day. It wasn't difficult to borrow the necessary gear, and a day later, I squeezed into the modern-day torture device known as a whitewater "play-boat," a kayak designed for a person with no feet and double-jointed hips.

Before I had time to complain, I was approaching the first ominous horizon line on the river, occasional explosions of mist hinting at what lay ahead. The river didn't disappoint, but the sheer volume of water was staggering. Massive islands created multiple channels, each of which would qualify as a big-volume river. Every river feature a kayaker could wish for was in abundance, as were many features that would give some of us nightmares. It was thrilling to paddle, but knowing that it would all be lost if the power companies had their way left me with a deep feeling of melancholy.

After several days of joyful paddling and untold liver damage, it was time to move on.

Once again, a hot tip from a fellow traveller suggested I couldn't leave Jinja without checking out "The Hairy Lemon." A tiny palm-fringed island in the middle of the Nile, surrounded by flowing waters and natural riverine forests, The Hairy Lemon attracted thrill seekers from around the world interested in one thing only – The Wave, also called "The Nile Special." This single wave tested the skills of the world's top freestyle kayakers, a hybrid sport resembling skateboarding crossed with surfing.[22] Luckily, I missed the crowd of adrenaline junkies, as I was there off-season.

After securing my bike amid thick foliage on the riverbank, I caught a ride to the island in a traditional *mokoro,* a long canoe carved from the trunk of a single tree. It was early afternoon when the local guide expertly poled his way through the swift currents and deposited me onto a hidden beach that led to the resort.

I was the only guest.

I had originally planned to spend two nights there, but it turned out to be almost impossible to leave. Besides, I had time. After one year on the road and nearly 50,000 kilometres of riding, I had reached the source of the Nile and experienced my first crash. I didn't feel too guilty about taking a few extra days off to rest and recover. Plus, I was intentionally slowing things down, hoping to wait out the rains in the south. As good as the main roads were in Uganda, everything unsurfaced would deteriorate quickly after even the lightest rain. And the wet season was in full swing.

The roar of nearby rapids filled the air, competing with the occasional rainstorm, and the tin roof above the lodge's central lounge amplified the atmosphere as bursting drops played percussion to a frantic rhythm. Then, as suddenly as it started, the rain ceased, and the sun emerged to slowly dry the saturated earth. This was a fertile land – they say if you spit on the ground, something will grow.

Red-tailed monkeys leapt gracefully between branches, foraging for berries amid the surrounding trees, while the noble African Fish Eagle looked down from its perch, keenly eyeing the river for its next potential meal. Countless species of colourful butterflies fluttered between shade and sunlight. A large tortoise meandered its way through the lush, green grass, pausing occasionally to nibble on a particularly juicy stem before pondering its next move. A monitor lizard, over a metre long, basked on a log in the heat of the sun, and two bright emerald-green snakes slithered between the rough-cut rafters above my head, searching every nook and cranny for snoozing geckos. It was such a stark contrast to where I began my journey.

The real world seemed so far away, if it existed at all anymore . . .

CHAPTER TWENTY-EIGHT

Not as Young as I Thought I Was

Personal Diary Entry: 11th June 2014

Day 394 (of 549)

50,038 km of 62,840 km (31,092 mi of 39,047 mi)

It is pleasantly rewarding to be this free, to go with the flow, to ride the wave, and let the voyage wash over me, bathing in the moment and savouring each experience.

Without expectations, it is impossible to become disappointed.

After four wonderful days at The Hairy Lemon Lodge on the White Nile, I reluctantly returned to my bike and loaded my gear for the short ride to Uganda's capital, Kampala. While visiting friends in Jinja, I was introduced to Kat, an inspiring teacher from England who devoted her career to improving schools in this part of Africa. She insisted I stop by on the way through Kampala so she could show me a little of her adopted city at a deeper level than that of a tourist.

It turned out to be one of the highlights of the trip.

Although an outsider, the locals welcomed me with overwhelming hospitality. Immediately, I was immersed in a world of some of Kampala's more interesting individuals. Dinner with one of Uganda's top reggae artists at a beautiful house overlooking the city was followed by cocktails in one of Idi Amin's former palatial residences on the shores of Lake Victoria. The expansive gardens were rumoured to contain macabre remnants from Amin's brutal reign. Nicknamed the "Butcher of Uganda," his rule was marked by brutal repression, ethnic persecution, and human rights abuses, leading to the deaths of an estimated 300,000 people.[23]

I had planned to spend a few days in the city with my new friend, after which I would continue to travel north towards Murchison Falls National Park. But over several drinks on a humid night, Kat decided she was ready for a road trip of her own. The following morning, I found secure parking for my motorcycle, and we loaded up her compact four-wheel-drive car and hit the road together.

Murchison Falls, on the Nile River, is several hours north of Kampala. It is the narrowest of many cataracts on the world's longest waterway, where the river is reluctantly forced through a gap less than six metres wide. The huge volume of water responds with a fury that is awe-inspiring and chilling. As a kayaker, I found it impossible to gaze upon such raw energy and not imagine myself in the thick of it.

The earth and air trembled.

Not even the clinging humidity could prevent a shiver of fear as I predicted the outcome of each possible approach, simple yet final. It was, without question, unsurvivable. As with fire, I find it hypnotic to look upon a river while losing myself in deep contemplation.

A lively backpackers hostel downstream provided a comfortable night's rest aided by cold beer and locally grown, fresh food. Kat and I couldn't resist viewing the falls from a different perspective, so the following day, we boarded a small boat for a gentle cruise up to the turbulent pools below the narrow cataract. Pods of hippos competed for riverfront access with solitary crocodiles while herds of elephants gathered to graze and bathe in the cool waters.

From Murchison Falls, we continued north along the dusty back roads through small towns and villages where children greeted our arrival with warm smiles and friendly waves. We camped further upstream by the equally impressive Karuma Falls, where the river splits into three channels to cascade over a series of steep drops. As we set up our separate tents to the sound of the powerful roar of angry white water, a local stopped by to suggest we keep a fire going through the night to dissuade any elephants from coming too close to our camp. A nearby village provided us with an evening's entertainment as we sought out a place to eat.

Our arrival sparked a flurry of excitement.

Children came to stare and smile at the strangers invading their sleepy town, giggling at our bizarre clothes and behaviours. Reaching out to shake a hand would result in screams and laughter as they ran to a safe distance before slowly allowing their curiosity to prompt a return to within arm's reach, where the whole process would repeat itself. The adults made polite enquiries about our purpose, confused about why we would come all the way here to simply look at a waterfall.

At some point during the dark evening, an elephant wandered into the village, and the locals erupted into action, banging pots and pans to scare off the massive creature. I briefly joined in the chase only to see the wrinkled, grey rump of the imposing beast disappear into the pitch-dark jungle. What followed were tales of constant harassment and adversity as the villagers struggled to protect their crops from these ravenous intruders.

That region of Uganda had been plagued by another, more sinister foe – the Lord's Resistance Army (LRA). Joseph Kony's band of vicious thugs terrorised the lives of so many. Several locals displayed the scars of bullet holes and knife wounds, permanent reminders of a cruel chapter in their lives, which began with their abduction as children and ended with their eventual escape.

***

From Karuma, we turned south along the main highway to Kampala, a pitiful strip of crumbling asphalt through dense forest. Speeds were controlled by deep potholes and savage speed bumps, some so steep we had to approach them at an angle to avoid scraping the underside of our vehicle. It felt strange to be behind the wheel of a car, but air conditioning and good tunes made it bearable.

Something someone said to me years ago kept circulating around my mind: "Driving a car is like watching a movie, while riding a motorcycle is like being a part of it."

Towards the end of the day, we stopped at a serene rhino sanctuary, camping within the confines of a strong steel fence. We had planned to take a guided walk the following morning to see some of the local residents, but as the sun set, three of the prehistoric-looking beasts wandered right up to our camp, intently grazing on the lush grass, oblivious to our presence. For an hour, we stood within ten metres of these incredible creatures – until a large, jealous male entered the scene.

The grazing stopped, and the tension rose.

Then the new arrival displayed a surprising burst of speed, chasing off one of the younger males, the noisy pursuit continuing long after we had lost sight of them in the dense vegetation. We assumed the show was over until we noticed a large female nearby, nursing a young calf. A heavily armed park ranger could name each rhino and give us a brief synopsis of their habits and personalities. The boisterous young calf's name was *Uhuru*, which translates to "Freedom" or "Independence" from Swahili.

Since I began this journey, I'd often been asked if I had a name for my bike.

Until then, I'd struggled to think of anything suitable. I had considered "Rocinante." As Don Quixote's horse, Rocinante was also his double. Like Don Quixote, I was definitely clumsy, past my prime, and well outside my area of expertise and capability, but my noble steed certainly wasn't. So Rocinante didn't seem like a good fit. The name "Uhuru" suddenly felt appropriate, and I vowed to finally christen my bike when we were reunited.

After six days on the road, we returned to Kampala in time for yet another dinner party. By sunrise the following day, Kat and my rafting friends had convinced me to change my route through Uganda and return to Kenya to attend their friend's wedding on the east coast. The sincerity and benevolence of those around me made the decision an easy one, so after a couple more days in Kampala, I loaded Uhuru, and began the long drive to the East Coast of Africa.

I stopped in Nairobi again to break up the journey, as the hospitality of Jungle Junction kept drawing me back in. Several other overlanders were camped there. I spent the day chatting with people and relearning the few Swahili phrases I had picked up on my previous visit. From our conversations, it was clear the recent bombings in Nairobi and Mombasa were still negatively impacting local tourism. Most resorts were empty, prices had plummeted, and too many people reliant on the tourism industry had lost their jobs. I picked up as much useful information as I could about the road I would be taking to the coast, traded contact information, and got an early start, anxious to cover the last 500 kilometres on one of the busiest roads in Africa.

Most of the imports and exports to and from this region of Africa went through the port city of Mombasa, with few choices of roads approaching the sprawling settlement. As the road dropped from the plateau of the central highlands, the temperature and humidity rose. It was a well-surfaced route, but the speed bumps and endless traffic made for a stressful ride. Heavy trucks and speeding coaches regularly pulled into the oncoming traffic where size trumped all on-comers and motorcyclists enjoyed few privileges. I lost count of the number of times I was forced onto the hard shoulder with mere inches between my handlebars and the oncoming traffic.

It was a tense ride with little opportunity to enjoy the spectacular scenery.

The drop in elevation brought with it a change in vegetation. Immense baobab trees dominated the terrain, their oversized trunks topped with disproportionately small branches, as though nature had briefly lost its sense of scale.

Just when I thought I was making good time, I felt the back end of the bike begin to wallow and weave. On one of the forced departures from my lane, I had picked up an acacia thorn sharp enough to

243

penetrate my rear tyre and the tube within. It took just over an hour, my fastest time yet, to patch the hole and get back on the road.

By early evening, I pulled into Mombasa, a city in a state of constant gridlock, not helped by the countless police roadblocks set up to prevent another terrorist attack. It didn't take long to find a backpackers hostel, and although it had been a long day in the saddle, I was energised by my new surroundings and keen to explore.

As a busy port in an impoverished country, Mombasa attracted more than its fair share of sex workers willing to risk their lives in this HIV hotspot for the price of a few beers. It was hard to have a quiet drink without being surrounded by prostitutes eager to earn a few extra dollars from this gullible white man. The attention could be flattering, but it often grew tiresome. I struggled to be as firm as I needed to be to turn down their advances, and they took that to mean they could wear me down if they persisted.

Eventually, I developed a strategy for dealing with their behaviour. I would ask one of them what she would make on a normal evening. I was often shocked to learn how little it was. For the same amount, I would ask her to sit with me, keep the other women away, and just have a conversation. No expectations, no sex. It made for a much more enlightening evening; the constant harassment would cease, and I'd get to know a new person, gaining an insight into a life I could only imagine.

***

From Mombasa, I took the coastal road north to the quiet town of Kilifi, checking into a beautiful backpackers hostel on the shores of a pristine lagoon.

The water of this particular lagoon supported a unique type of plankton that, when disturbed, would glow in the most magical display of phosphorescence, each wave creating a sparkling light show. Swimming beneath the surface after dark was like soaring through a swarm of fireflies while wading in the shallows created a ghostly trail of glowing, green bioluminescence.

I spent two nights enjoying one of nature's most beautiful spectacles.

But before long, it was time to join in the wedding celebrations – after a brief stop at a local market to pick up a pair of smart, if ill-fitting, trousers. Once the festivities began, any reservations about being an outsider imposing on the wedding of a couple I barely knew were quickly swept aside. Kenyan hospitality came easily, with no strings attached, and I soon enjoyed a party of epic proportions among people who made me feel like I was one of their own.

The beautiful ceremony was followed by music and dance that lasted well into the early hours of the next day, but it wasn't to end there. The party simply moved to a new location, and the revelry continued – until I realised I was not as young as I thought I was and retired to catch up on some much-needed sleep. After an eventful and deeply enjoyable few days, I was thoroughly exhausted.

But it was a good kind of tired.

CHAPTER TWENTY-NINE

The Man Who Gave Me My Name

Personal Diary Entry: 3rd July 2014

Day 416 (of 549)

51,518 km of 62,840 km (32,012 mi of 39,047 mi)

Africa is like no other place I know. It is a land of extremes: wealth, poverty, climate, terrain; yet the people who inhabit this vast territory are some of the most resourceful, generous, and joyful individuals I have ever encountered. It is a truly humbling experience to have those with so little share so much. It is often thought of as the poor continent, but I have found it rich beyond measure.

While I was restless, the lure of the peaceful coastline proved too seductive to resist. I was drawn back to the same quiet backpackers hostel outside the small town of Kilifi.

After yet another surreal night of swimming among the sparkling bioluminescence at a nearby deserted beach, I woke to a magical morning. A gentle breeze caressed the vivid purple flowers of the jacaranda trees overhead. The atmosphere was warm and humid, and the distant melody of reggae music floated through the air. Warm rains had saturated the earth, and the surrounding forest was bursting with life. I had discovered a secluded slice of paradise.

Had I stopped anywhere else, I would have moved on many days earlier, but this region of the Kenyan coast had an allure strong enough to suppress my wanderlust – for now.

My comfortable retreat was far from the busy roads and the pathological drivers on the shores of a large lagoon by the Indian Ocean. Thatched cottages nestled amid the dense jungle on a hillside overlooking the azure waters of Kilifi Bay. The natural surroundings, fresh air, and distant hiss of the ocean waves made for the perfect place to challenge my lingering feelings of disconnection.

It had been a long time since I had felt so complete.

While staying by the coast, I had ample opportunities to play with some of the most talented musicians in the region, jamming late into the night at the bar or on the beach. Bringing a guitar on this trip had been one of my best decisions, but I was still a beginner with lots to learn. The more I knew, the more I realised I didn't know, but my path was about more than arriving at a destination. It was about my incredible experiences along the way, and Kenya provided many of them.

The people I met and the energy I picked up on during my extended stay at the Distant Relatives Ecolodge made the entire journey worthwhile. The location attracted an eclectic group of travellers and explorers from around the world.

One particularly inspiring individual I met there was a fascinating woman with autism from Kenya. Driven by a desire to spread awareness of disabilities throughout the continent, she had embarked upon a walk. Not just any walk – she planned to cover the entire continent with two camels and a positive attitude. Disabilities within rural villages were often attributed to witchcraft, and those afflicted were sometimes excluded from their communities. In a bid to enlighten and inform, she was determined to show what a person with autism was capable of. While I knew it would take her many years to complete her crusade, she had already walked extensively throughout Kenya, and with her contagious enthusiasm and drive, I doubted anything would stop her from achieving her goal.

Also in residence was the determined group of young men and women building a traditional *dhow* further along the coast, intending to sail it around the world. Composed of a talented mix of nationalities, they planned to follow traditional trading routes while offering their skills and services to all they encountered. Their numbers included several medical professionals, engineers, teachers, and horticulturists. I was flattered when they offered me a position on their crew – but unfortunately, I had to decline to complete my adventure.

Anna, the quiet French architect, devoted her career to addressing the housing crisis facing the world's poorest. Based in Nairobi, she was working tirelessly to find a solution for the severe lack of infrastructure within the slums of Kibera. Close to 800,000 people lived there without access to clean water, electricity, a sewage system, adequate healthcare facilities, and established schools. While she could have been overwhelmed by the challenge of adding these services to provide a decent standard of living after the creation of the settlement, she was determined to try.

There was also a talented young German photographer living in the shanty towns of Kampala who was dedicated to capturing the essence of life in the faces of its residents. The portraits of the individuals she photographed told their stories perfectly. Her mastery of timing and lighting caught all the subtle details of the humanity in each person; it was almost possible to hear their voices and get a sense of their joy while feeling the weight of their sorrows.

Those were just a few of the amazing individuals with whom my path crossed. And then there was my companion whom I brought along with me, in spirit: Hendri Coetzee, "the greatest African explorer you have never heard of."

This was the man who gave me my name.

Hendri and I worked as guides on the Zambezi River many years earlier. He was the first person to refer to me as "Irish," a nickname that has stuck with me ever since. When we worked together, there were a total of five Daves working in and around the Zambezi River:

Scottish Dave, Kiwi Dave, Ozzie Dave, English Dave, and me – Irish Dave. Eventually, thanks to Hendri, I became known simply as "Irish."

Hendri and I shared many good times, but as our lives drifted apart, I lost touch with my old friend. I would often arrive at a river only to hear he had just left the area, but I was sure we would meet again at some point. Many years later, I unexpectedly stumbled across his obituary in an adventure magazine while I was living in California. It initially came as a shock to hear of his death, and I hoped it was not the Hendri I'd once known.

But deep down, I knew it could only be him.

As word of his death slowly filtered down through the news channels, the details of his final expedition became clearer. Snatched from his kayak by a large crocodile on the Lukuga River in the Democratic Republic of Congo, at age thirty-five, Hendri had achieved more in his short life than many could hope to in several lifetimes.

Shortly before his death, Hendri had started writing a book.

When *Living the Best Day Ever* was posthumously published in 2013, I was determined to pick up a copy so I could read about all of his remarkable accomplishments. I finally located the book among the river community in Jinja, Uganda, but I'd had little time to give it the attention it deserved. Only in Kilifi did I have the chance to sit peacefully and listen to Hendri's voice again. I'd read many tales of adventures in foreign lands, but his book stood apart from the crowd.

What I'd thought would be the standard fare of lists of achievements turned out to be beautifully written and insightful. Beginning as an account of a daredevil descent of the Nile River, from source to sea, it gently evolved into the story of a much deeper journey into what was behind his desire to explore Africa's most remote regions. As he matured, so did his prowess at capturing more than just the moment, delving deep into his motivations, and discovering that the only true adventure lies within all of us.[24] I assumed the book affected me because of the nature of my travels and the brief friendship I had shared with Hendri, but I went on to meet several people outside of our sphere who had been equally as touched.

***

Not far from Kilifi was the Tsavo East National Park. The sparse vegetation of grasses and thorny acacia bushes was dominated by massive African baobab trees randomly dispersed throughout the park, their thick trunks sharply contrasting their comical, spindly branches. As Kenya's largest park, it was home to a diverse selection of wildlife, none more impressive than the red elephants of Tsavo. These noble behemoths coated their hides with red mud to protect them from heat and insects and had long been a popular tourist attraction. Embedded deep within the park, a dedicated film crew had spent years patiently gathering footage of these enormous creatures.

It was within this region that the largest elephant in the world, Satao, roamed freely across the plains – until recently.

I was fortunate enough to meet several members of the film crew and listen to their stories of life within the park. I had been ignorant of the story of Satao, but the account of his savage slaying at the hands of ruthless poachers needs to be shared. Africa had thrown open its doors to foreign investment, selling off its natural resources at an alarming rate. No single state was pursuing these resources more aggressively than China. I had seen first-hand evidence of this tragedy as I moved south through the continent. Infrastructure improvements were underway everywhere to facilitate a more efficient resource extraction.

The Chinese government often imported labour, and locals I spoke to believed they frequently used convicts for the more physically demanding work, sometimes abandoning them overseas when a project had finished. The Far East has long been obsessed with a false belief that ivory brings good luck, and Africa has struggled for decades with the plague of poaching. All these factors, combined with modern GPS-tracking devices, night-vision technology, and automatic weapons, had left Africa's wildlife exposed and vulnerable, none more so than its elephants. Official statistics claimed that approximately ninety elephants a year were butchered for ivory in Tsavo, but it was widely believed the actual numbers may be ten times that amount.[25]

Collusion, corruption, and apathy contributed to the problem.

I had only seen pictures of Satao as he towered above his companions, his dark skin contrasting with his colossal white tusks. Those who worked closely with the animal believed he was conscious that his proudest feature had made him a target. He would often hide them amid the sparse bush.

In the end, it was a simple arrow fired from a primitive bow that took Satao down. Coated with a potent poison, he likely suffered a slow and agonising death. One can only hope he was dead before the poachers cut off his entire face with chainsaws. In the last picture I saw of Satao, he was lying alone, faceless, legs splayed, coated in vulture droppings amid the red dirt of the Tsavo plains.

The coastal region I was staying in also witnessed some disturbing events unfold.

A group of gunmen had hijacked two vans before driving into the nearby village of Mpeketoni, where they systematically murdered approximately sixty people. Initial reports suggested it was the Somali Islamist group al-Shabab. However, since then, the Kenyan police arrested the Governor of the region, adding to local suspicion that the attack originated closer to home. The attackers appeared to have targeted members of the ethnic Kikuyu tribe, the same tribe the president of Kenya belonged to. The attack, which began in the evening, lasted well into the night before the gunmen moved on to another village.[26]

In Mombasa, to the south, there had been more disturbances, targeted assassinations, and riots as the security forces wrestled to control the situation. Growing up in Northern Ireland had left me somewhat sceptical of the official reports issued by the government, and most of the locals seemed to agree. Civil unrest and fear can be powerful motivators when governments wish to impose more draconian laws on their people.

Yes, this region came with a unique set of difficulties. Still, it attracted a host of interesting individuals who were insightful, inspiring, and enlightening. The nature of my voyage often saw me drifting through countries, briefly meeting new people, and always being drawn further south towards my intended goal of reaching the end of the continent.

In Kenya, the realisation began to dawn that, for me, what makes life truly worthwhile is genuine, meaningful human connection.

I knew I would have to move on eventually, but I suddenly felt so content, surrounded by good people in one of the most beautiful locations on Earth. With my bike taking a well-deserved break and my physical progress temporarily suspended, I allowed myself time to just breathe.

I recalled how, when Sarah and I were preparing for this trip, we each said that if we found a place along the way that felt perfect, we could always stop and settle down for a while. Back in Europe, she thought Switzerland might be the place. Here I was wondering if maybe the sleepy town of Kilifi was it.

Cork to Kilifi had a nice ring to it.

Rumours abounded of more hidden gems and cool locations along the coast, both north and south. As much as I wanted to stop and explore further, the draw of South Africa sat impatiently at the back of my mind, pulling me towards my original goal. While there was still much to see and do, the previous few weeks had taught me to open my intentions and let chance meetings and serendipitous encounters play a more influential role in how I proceeded.

What adventures did the universe have in store for me, and what further challenges awaited?

CHAPTER THIRTY

The Curse of the Copper Doughnut

Personal Diary Entry: 26th July 2014

Day 439 (of 549)

51,740 km of 62,840 km (32,150 mi of 39,047 mi)

I chose a different path, one that would take me to some of the most interesting parts of this world, revealing tantalising glimpses of its most beautiful secrets while bringing me into the circles of people who are nothing short of inspirational, individuals who have chosen to live life by their own set of rules.

Because of the colour of my skin, I was allowed access to a privileged world, though it came at a small price. I am white. I am *mzungu*. I received preferential treatment over those who deserved it more than I did, and it was offered without resentment or question, as though I was somehow entitled to it.

The price I paid for this is what I called the "mzungu tax" – the cost of everything was inflated based on the assumption that I had more than most. It could get frustrating if I took it personally, but with a little patience and humour, we could always reach a fair compromise.

Travelling through Africa wasn't easy, but the rewards made it more than worthwhile. Each day was an adventure, and the attitude I approached it with made all the difference. An expression commonly heard here was "T.I.A.," or "This is Africa," used when things didn't go exactly to plan, usually accompanied by a resigned shrug of the shoulders.

After planning to spend just two days in Kilifi, the quiet town to the north of Mombasa, I was shocked to discover I had been there for almost two and a half weeks. The place itself and the people it attracted proved nearly impossible to separate myself from.

Each time I selected a day for when I would have to move on, I would make excuses as my chosen departure date came closer. I interpreted a series of stormy mornings with torrential rains and fallen trees as signs that the universe wanted me to stay.

But the wave of momentum that brought me to the east coast of Kenya had receded, leaving me fighting a growing sense of inertia, unwilling to embark on the next leg of this long voyage, torn between the unknown and the known, where I was comfortable. The friends I arrived with had long since departed to return to their lives elsewhere. Eventually, I decided it would be best for me to leave, too, before the loneliness settled in and I had no other company than my own.

I had almost forgotten how to load my bike.

Each piece of gear had its place, fitting together like a game of Tetris without room for anything extra. Picking up anything new along the way meant something old had to be left behind. Other than photographs and my journal, I knew there would be few souvenirs from this trip. After a couple of hours of packing, repacking, adjusting, and re-adjusting, I was finally ready to leave.

As I took the slick mud road from the backpackers hostel out to the main coastal highway, I was intentionally looking for a reason to return, almost hoping for engine trouble or a barrier in my path. But the road was smooth, and my bike seemed to miss being ridden. Uhuru and I gently made progress south.

I hadn't set my sights too far for that first day.

Soon, I was cruising through the bustling port city of Mombasa, then across the bay on a heavily laden ferry and through the dense mangrove swamps that dotted the coast. I then continued driving another hour to the coastal resort town of Diani Beach, a long stretch of white sand hugged by lush forest. A wonderfully eccentric couple I'd met at the wedding several weeks earlier had insisted I stop by on the way through their hometown. I pulled into a unique backpackers hostel where all the cabins stood on tall stilts.

There, I reconnected with another friend who had been staying in Kilifi during my time there: Anna, the architect based in Nairobi. Originally from France, her reserved personality added an air of intrigue to her stunning appearance. We had become close during our previous encounter, sharing stories and lessons from our journeys. We seemed to understand each other, and I valued our deep and meaningful conversations.

On a sunny afternoon, we sat on the pristine beach as Anna quietly shuffled a tarot deck.

"Will you do a reading for me?" I asked her.

"Do you believe in tarot?" she replied, giving me a quizzical look.

"Maybe." I hid the scepticism beneath my curiosity. "There's a lot in this world that doesn't seem to make sense. I'll try anything twice. What's the worst that can happen?"

"I don't think you are in the right headspace for a reading," her voice became serious as her deep brown eyes pierced mine. She was well aware of my recent history.

"Please!" I turned my body to face her, hoping she would see I was being earnest.

"Okay," she agreed, "but I'm not certain this is a good idea."

As she shuffled the deck, she had me cut it several times, and I marvelled at her dexterity as she manipulated the cards. They seemed to flow like water between her fingertips until the moment she abruptly stopped and declared they were ready.

"We'll start with a simple spread," she began. Laying the deck in front of me, she had me choose three cards, which I set face down in the space between us.

"Have you set your intention?" she asked.

"My what?"

"Your intention."

I could sense a little impatience in her voice.

"Before we begin, you should set an intention, like a question you want answered or an issue you want resolved. The cards may not provide the solution, but they might give you some guidance."

A cascade of questions flooded my mind. There were so many things I was uncertain about – endless unknowns about my past and future. It took some time to wade through them all and find the most pressing concern that troubled me: *Should I continue on this journey? Am I on the right path?*

"Are you ready?" Anna asked at the exact moment I had made my decision. She seemed to have read my mind. "Now, turn over your first card."

What happened next still haunts me.

I heard her draw a sudden breath as I turned the card, and I looked at her for reassurance.

"Are you sure you want to proceed?"

"What does it mean?" I asked.

"It's too soon to tell. The cards . . . they relate to each other. I just want to know if you are ready for something you may not want to hear." The nervousness in her voice was unsettling.

I turned the next one and watched her eyes shift rapidly between the faces of both upturned cards. When I turned over the final card, I watched the colour drain from her face as she covered her mouth and gasped.

"We need to stop!" she cried out.

In the short time I had known her, I had never seen her so visibly upset, and it frightened me. "What is it?" I pleaded.

"This was a bad idea." She wouldn't look me in the eye. "You were not ready for this. What was I thinking? I'm sorry. I am so sorry." Her voice trembled, and I noticed her hands were shaking as she gathered the cards. After collecting the deck, she quickly returned it to its ornate box before standing to leave.

"Anna, what's wrong? What did you see?" My initial scepticism was giving way to an uneasy feeling of apprehension because of her tense disposition. No matter what I said, I could not convince her to reveal what message the cards had relayed.

"I should go!" She hurriedly gathered her belongings. "I need to be alone."

I watched her in stunned silence as she walked away along the empty beach until she vanished from sight. When we met again later that evening, she had regained her composure, but she made it very clear – she did not want to talk about what had happened.

***

After a couple of days of enjoying the pristine local beaches, savouring exotic cocktails on precipitous cliff-tops overlooking the roaming elephants of Shimba Hills National Reserve, and swimming in the warm waters of the Indian Ocean, I decided to move on. Before I left the region, the couple I had met at the wedding, whom I had initially come to visit, extended an invitation to one last dinner party at the exquisite home of some friends who lived near Tiwi Beach.

As I parked my bike outside their place, I sensed something was wrong. The engine wasn't running as smoothly as normal. When I got ready to leave later that evening, the bike refused to start. A flat battery hinted at a charging problem.

My heart sank as I realised I was stranded yet again.

After a peaceful night in a cabin by the ocean, I was able to top up the battery with enough power to get me back into the town of Diani Beach, where I hoped to fully diagnose the issue. Several weeks earlier in Nairobi, I'd had the alternator repaired and suspected this might be the problem again. So, after settling into another backpackers hostel, I opened up the engine.

While a visual inspection seemed to confirm my suspicions, I removed part of the alternator and tested it just to be sure. A conductivity test verified the stator had developed an earth leak, so it would no longer provide a sufficient charge to the bike's battery. With several key components relying on an adequate electrical supply, my bike was essentially crippled.

The curse of the copper doughnut had struck again.

Repairing the same part a second time seemed like an exercise in futility, so I began the search for a new component. Multiple forums existed online from other riders having similar issues. Even though BMW was aware of the problem, they insisted on selling the entire alternator at an extortionate price when the fix required a single piece of the complete unit. Finding an aftermarket part required nothing more than a quick internet search.

Getting it into Kenya would prove to be somewhat more complicated.

Shipping accessories to that part of the world could often be a lengthy process, as Kenya's customs officials were always on the lookout for ways to extort additional "fees" on anything coming into the country. Other bikers had had their parts held up for weeks at the airport while dealing with corrupt bureaucrats. It seemed as though the only problems I had encountered along the way on my long voyage involved people in uniform, so I began a search for a suitable courier to bring in the part.

Several acquaintances suggested I contact a group of Somali "importers" who could bring absolutely anything into the country for a modest fee. Then, a close friend suggested I contact an online group of Kenyan expats who sometimes did favours for each other. Before long, I had the details of an especially helpful woman planning to return to Kenya from California. I had the part shipped to her, and within a week, it was on its way to Nairobi.

In the meantime, Chris from Jungle Junction had contacted me. He needed a motorcycle to be moved from the coast to the capital, so with a little planning, it was easy to coordinate riding that bike to Nairobi with the collection of my replacement part. I was a little nervous about riding an untested machine along the 500-kilometre road from Mombasa to Nairobi, so I spent some time in Diani Beach testing and adjusting the stranger's bike before setting off.

Being uninsured and riding a bike not registered in my name reminded me of my youth, dodging roadblocks and riding as inconspicuously as possible. Just like Northern Ireland, these heavily used Kenyan roads had more than their fair share of police checkpoints, a result of the heightened security because of several recent shootings. I employed my tried and trusted method of simply waving back at the police officers when they flagged me down. At one checkpoint, I observed an officer pulling out his mobile phone as I sped past. I imagined he was warning his colleagues down the road about my imminent arrival. At the next opportunity, I pulled over and enjoyed a relaxing lunch, hoping that they would have forgotten all about me by the time I resumed my journey.

I was reminded on several occasions that motorcycles were expected to get off the road when an oncoming truck or bus wished to overtake; I even witnessed one large coach with an impossibly heavy load lashed to its roof tilting over onto two wheels as the driver struggled to return to his lane, narrowly avoiding a head-on collision with an oncoming truck.

The white-knuckle ride went smoother than I expected.

While in Nairobi, I was also able to catch up with some good friends before embarking on the return journey via bus. Having seen first-hand how the bus drivers operated, I couldn't help but feel a little apprehensive as I boarded the night service back to the coast. By early evening, the dilapidated bus was making its way, slowly, through the traffic-choked streets of Nairobi.

Once we had cleared the outskirts of the city, the driver made every effort to push the vehicle to the absolute limits of its speed and manoeuvrability, swerving recklessly from behind overburdened trucks, sometimes completing a pass but, more often than not, wrestling the bus back to its original lane to avoid a collision. The coach company had advertised an onboard DVD-entertainment system, which, of course, was not working. However, nothing could compete with the drama unfolding beyond the cracked windshield: a winding, swerving snake of headlights and taillights resembling a fast-paced video game.

Without warning, the driver pulled over next to a nightclub and a brothel in a sleepy roadside village and announced that he was going to take the rest of the night off. The bus erupted with complaints from passengers who had been tricked into thinking, like me, that the overnight service would drive through the night. Amid all the confusion, nobody noticed that the driver had quietly slipped out of the vehicle and into the bar. By 5 o'clock the following morning, tempers had calmed, and the driver returned to continue the journey, resuming his frantic race for the coast interspersed with unexplained stops in the middle of nowhere.

At one point, we slowed to a crawl and gently eased into the adjacent lane. Through the side window, I noticed a lone shoe in the middle of the road, followed by a school notebook, its crisp white pages fluttering in the breeze. Then we passed a sombre line of villagers quietly standing along the edge of the asphalt, their gaze fixed upon the twisted remains of a young cyclist lying in a pool of the brightest red blood. Less than two minutes later, the bus accelerated to full speed, and we were haphazardly weaving between lanes again.

Eventually, I was reunited with Uhuru.

After more than a year in the saddle, my motorcycle had become more than a mere machine. It was my loyal companion, an extension of my very being. There were moments on the open road when I felt an almost mystical connection to its energy resonating through my body. Sometimes, I imagined the fluids coursing through its veins and mine were entwined, a shared lifeblood that bound us together in a symbiotic dance of motion and purpose. The rhythmic hum of the engine had become the heartbeat of our progress.

A journey that had taken me seven hours on the motorcycle took twenty-two hours using the local bus network. I was excited about returning to Diani Beach with my replacement motorcycle part, keen to fit it and get back on the road. I had been on the coast for almost two months, and Kenya, though fascinating and beautiful, was not the cheapest country in Africa. My new stator, supplied by the Electrosport company of California, was reported to have a higher design specification than the original and be less prone to overheating. I spent a day with the bike, fitting the new part and replacing the oil and coolant while performing other routine maintenance procedures and testing the new part. It appeared to perform well, and I readied myself to turn south yet again.

Diani Beach had made for a fine place to get stranded, and I could not have wished for a more interesting group of people to spend my time with. Home to one of the most beautiful beaches in the world, it attracted a diverse group of skydivers, kiteboarders, divers, and travellers from around the world. The expat community was booming, and those there for just a seasonal gig also added to the diversity of the place.

The locals were even more entertaining.

I had lost count of the gatherings I'd attended since my arrival, constantly being invited to join someone somewhere. Those people knew how to live life – always celebrating something or making up an excuse to celebrate the mundane.

I'd seen more sunrises in a single month than in many previous years – not necessarily intentionally. Kenya is probably where the expression "time flies when you're having fun" originated from.

Everyone planned to lose track of time – on purpose.

Occasionally, I'd meet those who were more serious about how they spent their time. A couple I met in Nairobi asked me when I planned to settle down, get a real life, and make plans for my retirement. It left me pondering the definition of a "real life." If it was the career, the house, the car, the wife, the children, the possessions, then I wanted to choose a different path. It was neither right nor wrong, real nor fake. It may be seen as unconventional, but it was what made me happy. I knew how it all began, and I felt pretty certain I knew how it would all end, but I felt fortunate enough to get to decide what the content would be.

I planned to make it as interesting and enjoyable as possible.

CHAPTER THIRTY-ONE

My Dirt Was Authentic

Personal Diary Entry: 14th August 2014

Day 458 (of 549)

53,221 km of 62,840 km (33,070 mi of 39,047 mi)

I feel free. Free from the self-imposed confinement we generally consider to be the only possible way of life. Free from the debts and obligations that would chain me to an existence that is often thought of as "normal." Free from the possessions that would tie me to a location by their very being.

With a renewed sense of purpose and a fully functional bike, I could finally drag myself away from Diani Beach. After my extended stay on the coast of Kenya, it felt like I was leaving behind a little piece of myself. I had met so many interesting people and made some lasting friendships while in this region that it was beginning to feel like home. In the end, just like in Kilifi, I had three "last nights." Each time I renewed my stay, the locals assured me that I could check out any time I liked, but I could never leave.

Tanzania had felt so distant when my bike had been out of action, but it took little more than an hour to reach the border along the scenic coastal road. The crossing went smoothly, but the Kenyan customs officials couldn't resist trying to squeeze a few more shillings out of me as I completed my exit paperwork. Apparently, I hadn't paid my "road user fees," and they needed a modest contribution before I received the final exit stamp. I explained to them that I hadn't been using their roads since my bike had been crippled by a botched repair job done in Nairobi. Obstinate border officials had constantly been a headache throughout the trip, but I'd always gotten through eventually. I had time on my side, and as a long line of impatient truckers gathered behind me, the customs official finally relented. I was on my way once again.

Entering Tanzania was relatively straightforward. As I approached the border, a swarm of money changers and fixers surrounded my bike, promising impossibly good rates and invaluable help, but, this time, their assistance was unnecessary. I got through all of the posts without a hitch.

I unleashed Uhuru, and we were soon coasting down Tanzanian roads, which were noticeably better than those of Kenya. Gone were the gaping potholes and the vicious speed bumps that had left me airborne on more than one occasion.

Still, I was extra careful.

Each time an oncoming automobile would steer into my lane in an attempt to pass a slower vehicle, they would timidly return to their lane on seeing my bike where they had thought there was vacant space. As a reflex, I instinctively swerved towards the edge of the road every time this occurred – just in case.

By early afternoon, I was pulling into a rundown hotel in the dying seaport of Tanga. The once-busy port struggled to compete with nearby Mombasa, and the town appeared to be a slowly fading relic with its best times behind it. Early the following morning, I was back on the road and following the coast south to the small town of Bagamoyo. Once the capital of German East Africa in the late 19th century, it had been in an extended state of decline ever since, with the crumbling town centre revealing hints of a more decadent past.

I felt quite liberated to be in a less restrictive country.

Kenya's reputation for crime had been one of its less appealing traits, and Tanzania had a much more relaxed feel. I found a comfortable camp by the beach and set about exploring the town. Walking through the streets in Kenya after dark would have been considered reckless, so it felt good to finally let my guard down a little. As night fell, I went for a peaceful stroll along the water's edge, dipping my feet in the warm waters of the Indian Ocean.

Not far from the shoreline, the orange glow of multiple fires drew my attention. As I got closer, I discovered a lively fisherman's market where all sorts of unidentifiable sea creatures were being cooked over endless rows of glowing charcoal fire pits. Sweating locals tended large cast-iron pots bubbling with oil. The heat and smells were oppressive, but the atmosphere felt authentic – a beautiful snapshot of real Africa.

I found a busy restaurant nearby, frequented by a mixture of locals and foreigners. A kind group of volunteer teachers from Denmark noticed I was alone and invited me to join their table. Once I was seated, they began telling me all about their friends who had just been mugged on the same beach that evening by a group of machete-wielding thieves. When I asked them about other areas in town that were known trouble spots, they told me to avoid the fisherman's market at all costs.

So much for my well-honed instincts.

As I wandered through the quiet streets on the way back to camp, I was surprised to see faint glimmers of light emanating from many of the buildings I had wrongly assumed to be derelict. I had likely come too close to people's homes this way, thinking I wasn't trespassing on anyone's property. The entire day proved to be a good reminder that I didn't know everything. Africa was a different world – and anything could happen.

I spent two days in Bagamoyo this way, wandering through the neighbourhoods and eating fresh seafood at every restaurant with an inviting atmosphere. During that time, I struggled with deciding whether to visit the nearby island of Zanzibar. I eventually opted to turn inland. My extended stay in Kenya had left me only two months to reach Cape Town, and I still had such a long way to go and so much more to see.

The road took me west and up into higher elevations, and it was refreshing to feel the temperature dropping. I stubbornly insisted on always wearing my protective riding gear for safety. But it became quite uncomfortable in the hotter regions. Much of where I'd been in Africa so far included humid jungles, sunny beaches, or blazing hot deserts.

As I peacefully rode along, I was pleasantly surprised to find plenty of effective road signage, making navigation relatively easy.

Before long, I was cruising along a quiet highway through Mikumi National Park keeping to the imposed speed limit of seventy kilometres per hour – a reasonable pace for somewhere that doesn't have abundant wildlife gathering near the road. Here, I was rubber-necking the entire way as elephants, giraffes, impala, zebra, and baboons all made an appearance, making it hard to keep my eyes on the road ahead. I found a comfortable campsite just outside the park and settled in for a quiet night as a blood-red sun set over the endless savanna.

The next morning, I rose with the sun and kept travelling.

While making good time on my journey inland, I noticed that each town I passed through had at least one group of police officers in crisp white uniforms, sheltering from the intense sun in the shade of the roadside trees. Many overlanders coming from Tanzania had warned me about the ruthless and plentiful speed traps. Fifty kilometre-per-hour speed limits in the towns and villages were well-posted before entering, but, more often than not, the signs indicating the end of the controlled zone were missing, leaving me confused as to whether or not I could wind back the throttle and return to a more reasonable speed.

The local police knew the exact location of these areas of confusion and, on one occasion, pulled me over for going a little too fast while leaving a small village. Unlike in Kenya, the checkpoints there used solid barriers, so I was unable to simply wave and drive past. With a smile and a bit of humour, I was able to talk my way out of a ticket. I continued my drive, stopping briefly in the small town of Iringa, where I stumbled across a quiet guesthouse staffed and managed entirely by deaf people from the local community.

After a silent, delicious lunch, I decided I still had time to get a few more kilometres behind me before I'd have to look for a place to stay, so I got back on the bike and kept moving. It was late afternoon when

I spotted a police officer blocking the road ahead, raising his hand in the air. I was sure I'd already left the speed limit zone, tempting me to high-five him as I rode past, but a police car parked just off the road made me change my mind at the last second. As I came to a stop, he informed me I'd been doing eighty-nine kilometres per hour in a fifty zone. He proudly pointed to the bright neon-red numbers displayed on his radar gun. With a broad grin, he announced that I must now pay the fine.

It was hot, I was tired, and the charm I'd relied upon earlier seemed to have no effect.

The officer directed me towards the police car and told me to talk to "the boss." Sweating profusely underneath a stained and strained uniform that struggled to contain his ample proportions, the police captain smiled at me from the passenger seat. Thick piles of Tanzanian currency were stacked neatly across the dashboard.

The bidding began at 100 USD, with his smug attitude implying he now held all the cards. His expression changed when I produced my "fake" wallet and showed him its meagre contents, supplemented with expired credit cards and an old driver's licence for an authentic look. After threatening me with jail and a court appearance, he finally agreed to settle for twenty dollars. When I asked for a receipt, he became quite upset, brusquely scribbling the details of my offence on a semi-official-looking piece of paper before sending me off with a dismissive wave.

As I mounted my bike, stuffing my first speeding ticket in over 50,000 kilometres into my now empty wallet, I resolved to spend less time in Tanzania than I had originally planned.

After two more days of hard riding, I was approaching the Malawian border.

Border formalities were painless, and for the first time in many months, I was allowed to enter a country without having to pay for a visa. I instantly warmed to the people of Malawi. It initially came as a shock to find so many people on the roads: men riding bicycles stacked high with supplies, women balancing enormous loads on their heads, children skilfully playing football with balls made from nothing more than plastic bags and string, very few other vehicles and, most importantly, no speed bumps. Police checkpoints were frequent, but they happily waved me through when they recognised that my bike was not local.

On a few occasions, I was stopped and asked the usual three questions:
"How fast?"
"How many CCs?"
"How much did it cost?"

By now, that brought my well-practised response of "240 kph, 800cc, and £10,000." At one point, a lightly armed soldier stepped into my path and flagged me down, only to give me a heartfelt welcome to Malawi and a firm handshake.

By early afternoon, I'd reached a camp recommended by another rider. Situated on the sandy shores of the turquoise waters of Lake Malawi, it felt like the perfect spot to spend a couple of nights. The appearance of several commercial overland trucks carrying intrepid tourists reminded me that I was now getting into the more-travelled regions of Africa. I spent evenings under the stars enjoying good company around blazing campfires.

Once more, I strummed my guitar and was humbled and impressed by other traveller musicians. Some people made do with so little – old ukuleles with missing strings or those without instruments simply slapping an improvised drum to the rhythm. Illuminated by firelight, we made beautiful music late into the night until, slowly, the group got smaller and smaller as we retired to our various sleeping arrangements one by one.

A few days later, another tip from a fellow traveller saw me turn off the main road into the nearby mountains, climbing steep, rough dirt tracks towards the quiet eco-lodge known as "The Mushroom Farm." I spent two nights at the cliff-top lodge overlooking the vast expanse of Lake Malawi, rising before dawn to catch sight of the fishermen's boats dotting the lake. Slowly, each one would make for shore to dry their catch before the horizon began glowing with the deep pinks and reds of the imminent sunrise. I'd become transfixed by the unfolding light show as the sun split the horizon, casting crimson reflections across the distant grey waters, completely absorbed by the moment.

I was convinced I'd stumbled across yet another little slice of paradise.

At the end of each day, as the sun set and darkness fell, one by one, the fishermen would take to the water again, casting their nets by the light of the oil lamps suspended from their dugout canoes. It looked like a line of fairy lights stretched across the visible horizon.

With each day that passed and every kilometre I covered, it felt as though I was nearing the end of my journey. Only a handful of countries now separated me from my final destination. As I encountered more travellers coming from the south, I was constantly reminded of how close I was getting to Cape Town.

I still needed to finalise my eventual approach – as there were many options – but the clock was ticking, and my budget was dwindling, so it would soon be time to make some hard decisions. The bike was performing well and my gear was holding up, but the signs of hard use in tough conditions were becoming evident on my equipment and me.

With luck, it would all last long enough to get me to the end.

As I mentally prepared to conclude this chapter of my life, I thought about how many different beginnings and almost as many endings I had experienced in my forty-two years on this planet. The beginning of my life. The beginning of a love story. The end of one. The beginning of my first long-distance motorcycle trip. The end of that trip. The beginning of my second even-longer-distance motorcycle trip. The impending end . . .

I thought back to a moment I experienced a few days back. From the sun-drenched shores of Lake Malawi, I observed a peculiar phenomenon unfold over the water. The air thickened with ethereal tendrils of what looked like smoke. Clouds of lake flies appeared to burst from the water's surface. Each swarm must have contained millions of the tiny creatures; they danced and spiralled with a hypnotic rhythm as though of one mind. It was the mating display of the short-lived adults; one last glorious performance before they completed their cycle of life; a reminder to embrace all the fleeting beauty that graces our brief time on this earth.

Would Africa give me one last glorious performance before my trip's end?

***

From the serene Mushroom Farm Eco-Lodge, high up on the Nyika Plateau overlooking Lake Malawi, I descended into the Great Rift Valley to follow the shoreline road south to Nkate Bay. I found a camp just outside the quiet fishing village and pitched my tent by the water's edge beneath the hot afternoon sun.

A refreshing swim in the crystal clear waters of Lake Malawi helped cool me down afterwards. Diving into unsalted waters was a strange sensation given that the lake was as vast as an ocean. When fully submerged, the water tasted sweet, and my eyes did not sting when I opened them. I could see colourful cichlid fishes darting between the rocks, their bright markings shimmering in the penetrating sunlight.

Above me, locals slipped silently past in their dugout canoes, gathering supplies for their nighttime fishing excursions. I took advantage of an afternoon boat ride to a nearby beach where fishermen dried their catch and mended their nets. Stopping along the way, I admired a fish eagle as it gracefully swooped across the surface of the calm waters to pluck out a fresh meal from the plentiful fish that populate the lake.

After a peaceful few days in the region, I packed my bike and continued south, stopping for a couple of nights at the isolated town of Kande Beach. The beach camp on the white sandy shoreline provided endless opportunities for exploring, and I couldn't resist swimming out to the nearby Kande Island. It provided me with a good reminder of just how unfit I'd become since starting my journey. Almost a kilometre from shore, there were a couple of occasions when I wondered if swimming that far had been such a good idea as the island seemed to get further away the more I swam towards it.

Tired but exhilarated, I eventually dragged myself onto the island's bouldery shore and spent some time resting on the warm slabs of rock, enjoying the serenity, totally alone – apart from the nesting fish eagles and shy iguanas. As I stared back across the expanse of open water, the stupidity of the predicament I had gotten myself into settled upon me. I hadn't told a single person about my intentions, and I had no idea what kind of predators inhabited the water.

It took a little time to summon the courage to swim back.

Later that evening, I rewarded my foolishness with a cold beer at a local bar. Charley Boorman's *Long Way Down* T-shirt was proudly displayed on one of the walls. When I was asked about details on my trip, I often heard the response: "That's just like *The Long Way Down!*" Aired in 2007, it was a popular television series in which Ewan McGregor and Charley Boorman took a motorcycle trip from London to Cape Town.[27] As I entered the more-travelled parts of Africa, I began to hear stories about how people remembered the actors' arrival in camps and villages along the way, preceded and followed by their entourage of fixers, medics, mechanics, camera crew, and assorted support vehicles.

The bar owner chuckled when he recalled how Ewan and Charley had ridden in years before, their bikes and gear caked in a fine red dust while the rest of their party appeared perfectly clean. The thing is, there is no such dust for many, many kilometres in this region. Along the shores of Lake Malawi, all the smaller roads consist of a fine white sand, which, while difficult to ride through, certainly did not create the same effect. He suspected a make-up and wardrobe crew had coated the lads in red dirt to make for a more dramatic entrance – so much for "reality" TV.

As much as my trip resembled theirs in terms of some of the routes I'd taken, it certainly lacked the sheer numbers of people and theatre involved. My journey danced to a different rhythm, a quieter, more intimate melody. I was okay with that. This adventure was uniquely mine.

My dirt was authentic.

It was hard to describe, but the feeling of throwing my leg over the bike each morning and having absolute freedom to do whatever I wanted, to chase my whims and wander wherever my heart desired, was such a liberating experience that I would have had it no other way. There were times when I felt exposed, vulnerable even. However, rather than dwell on the "what ifs," I focused only on the present. Every decision I had made in life had led me to this moment, and it would only be my attitude that defined how I reacted to and perceived each new experience. Each day was an opportunity to create my narrative.

In the simplicity of the open road, I was becoming more aware of the complexities of the self.

CHAPTER THIRTY-TWO

The Seed from Which Reverence Grows

Personal Diary Entry: 27th August 2014
Day 471 (of 549)
54,919 km of 62,840 km (34,125 mi of 39,047 mi)
Pulling into a camp on a loaded motorcycle often invites curious glances from fellow travellers and locals. Frequently, I'm approached by inquisitive onlookers who want to know about my journey and the bike. Reactions vary when I talk about where I've been and how long it has taken me to get this far.

From Kande Beach, I continued south towards Cape Maclear, a forested peninsula at the southern end of Lake Malawi, finally settling into a quirky little camp called Fat Monkey. With its unique aspect, the beach provided an idyllic location for observing sunrises and sunsets. It was the perfect place to enjoy what the locals referred to as a "sundowner" – any cocktail of your choice, in my case, a gin and tonic – savoured while watching the sun descend. The drink went down smoothly as the sky turned deeper shades of red, and the sun gently kissed the water's edge before being engulfed by the hazy horizon.

Is there anything more magical than the sun rising or setting over water?

After three days at Cape Maclear, I turned west and climbed out of the Rift Valley towards Malawi's capital, Lilongwe. Several months before, I'd contacted an old friend I had worked with on the Zambezi River, and she insisted I stop by if I was ever in her area. So after a beautiful twisting climb out of the valley, I was soon on the road to her house.

It had been a long time since I'd seen Juba, but she welcomed me as though it were yesterday. I'd planned on spending only a couple of nights in the city, but she talked me into staying a little longer, giving us plenty of time to reminisce about old times and catch up on all that had happened since.

After so long on the road, it was a truly wonderful experience to spend time with Juba and her beautiful family in their precious oasis of peace and tranquillity. They showed me around, taking me out on a tour of the city, a tidy metropolis of old and new districts divided by parks and wooded areas. Her generous husband introduced me to the game of golf – my first ever, which I subsequently won with an impossibly high score.

That's how it works, right?

With Juba's encouragement, I even gave a short presentation to several hundred children at her local school. Their eager faces showed signs of rapt attention and fascination as they took in the details of my story, and I hoped I inspired them to dream of one day embarking on their own adventures. Once upon a time, my journey was nothing more than a dream.

After a very restful stay in Lilongwe, I made my way towards the Zambian border, less than an hour away. At one final police checkpoint less than ten kilometres from the border, I was stopped by a tall, thin officer in a loosely fitted uniform. Before he had a chance to speak, I bombarded him with questions about directions, distances, and the nearest fuel stations. I could sense he was fishing for an opportunity to write a ticket or issue a fine, but before he had the chance, I began telling him how grateful I was that the police in Malawi weren't like those dirty cops in Tanzania and Kenya.

What could he do but heartily agree?

After exiting Malawi, I walked into the Zambian customs office and noticed multiple posters outlining the signs and symptoms of the Ebola virus. With a recent outbreak on the west coast of Africa, daily rumours were emerging regarding suspected cases throughout the continent. It looked as though Zambia's government was taking the threat seriously.

A large lady with a serious face informed me that she would have to take my temperature, and I audibly gulped at the prospect, wondering how she planned to do it. She then asked me to expose my chest before proceeding to point a laser thermometer at me. Suddenly, she frowned, looked at me, double-checked the numbers displayed, then tried a second time.

It was shortly after midday, the sun was high, the air was hot, and I was wearing my full riding outfit. My temperature was above average after walking between the various offices located around the border post to find the Ebola screening facility. What followed was a list of questions about my physical condition and my travel history.

Finally, I satisfied her that I was not infected with anything other than poor dress sense and an inappropriate sense of humour. Once again, I was pleasantly surprised to receive a visa at no cost for Irish citizens while travellers around me pushed fifty-dollar bills across the counter to the immigration officer. Back on the road, I turned north at the small farming town of Chipata towards South Luangwa National Park. By late afternoon, I was riding into Croc Valley Camp on the banks of the Luangwa River, stopping briefly to allow a herd of elephants to cross the road in front of me.

There were times when my mind would drift, and I'd briefly forget I was in Africa when the terrain reminded me of another time and place. But it would only take a second to return to the reality of where I was, when something uniquely African happened. Elephants disrupting traffic on a dusty, dirt road was one of those occasions.

As I pitched my tent, I could hear the loud guffaw of a hippo wallowing in the muddy river bed below, sounding like an evil villain in a low-budget horror movie. As night fell, the noises from the park appeared to rise in volume. The occasional roar of a lion interspersed with the yapping of hyenas mixed with more sinister sounds I could not identify. For once, I was thankful the river contained a healthy population of crocodiles, hoping it would deter the park's residents from wandering into my open camp.

I found it difficult to sleep while trying to identify the source and proximity of each sound. Sometime around 3 a.m., I heard the unmistakable sound of an elephant ripping branches from a neighbouring tree. It sounded close – *very* close – and my heart began to thump as I wondered what was best to do: stay in my tent or run for my life.

In hindsight, I probably did the dumbest thing I could have done.

I quietly unzipped the outer flysheet of my tent to get an idea of how close the animal was. At the same time, I was fumbling with my camera in the dark to find an appropriate night setting. I could see very little when I looked outside, but I could hear movement nearby. I pointed my camera in that direction. I hadn't expected the flash to go off.

Neither did the elephant.

The sudden burst of bright light illuminated its imposing form, walking straight towards my flimsy tent. I ducked my head back inside the tent, embarrassed by the little squeal that had somehow forced its way past my lips. My heart raced, and my mind worked overtime, planning my next move. I was trapped, and all I could do was wait. I readied my knife, hoping I would have time to cut my way out should my exit become blocked. It seemed like an eternity as I remained silent, listening to the soft footsteps approaching and expecting my tent to collapse at any second.

As the elephant passed by, it gently brushed against the thin nylon outer of my tent. Before I knew it, it had vanished. I waited for some time before crawling out of the tent to see if there was any damage. Everything seemed okay. My tent was still in one piece and my bike stood upright nearby. As I swept my powerful headlamp around the camp to assess the elephant's whereabouts, the bright beam fell on the squat, solid body of the largest hippo I had ever seen, keenly munching grass less than ten metres from my tent.

As I lit up its face, a vague memory flickered in the back of my sleep-deprived mind. Something like, "Never point your flashlight at a hippo unless you want to make it very angry."

It stopped chewing and gave me a look that seemed to imply, "Just try that again if you want trouble."

I instantly extinguished the light.

As the adrenaline wore off, I decided to go find a toilet before crawling back into my sleeping bag for a restless night's slumber.

The following day saw me up bright and early before collapsing, exhausted, into a hammock shortly after breakfast. In the afternoon, I joined a safari tour into the park. As with most other parks, motorcycles were prohibited, so I jumped onboard an open-topped Toyota Land Cruiser with a guide, driver, and several other tourists. It turned out to be a rather strange experience, although we saw many different species, including lions.

It all just felt a little contrived. At one point, we even got into a traffic jam in a park that stretched over several thousand square kilometres as each vehicle jockeyed for the best spot to observe a passing herd of buffalo.

At times on my trip, I felt like I was suffering from experience overload, taking for granted the wonders all around me. I have heard of that occurring during people's first visit to a wildlife park. Upon entering, one is enraptured by the smallest of things – impala skipping through the bush or warthogs wallowing in the mud, but, by the end, we give only a secondary glance to the woolly mammoths and the sparkling unicorns. I sometimes needed to remind myself of where I was and how fortunate I had been to have experienced all that I had.

As my journey was nearing its finale, I became more aware of the need to cherish every moment and see it all with the wide-eyed wonder of a newborn child. With this newfound perspective, I approached each day as though I was experiencing the world for the first time. I drank in the vibrant hues of each sunrise that painted the skies, marvelling at the delicate dance of light and shadow. I listened to the whispers in the breeze, deciphering the mysteries they revealed – were they telling me that rain was coming or that tomorrow would be even warmer than today? I realised the true beauty of life did not lie solely in the grand and the magnificent, but also in the quiet, hidden corners waiting to be discovered.

Appreciation is the seed from which reverence grows.

CHAPTER THIRTY-THREE

One Life Is All You Get

Personal Diary Entry: 17th September 2014

Day 492 (of 549)

56,356 km of 62,840 km (35,018 mi of 39,047 mi)

As I navigate the river of my existence, I find myself ruminating on the sombre truth that the debut of an experience is a solitary affair. I can encounter things for the first time, only once. There is an inherent yet tragic beauty in each of these inaugural moments, an ephemeral quality that makes them stand apart from the continuum of time.

Be it the first kiss, the first taste of an exotic flavour, the untraveled road or the unheard song, each becomes a shining star in the constellation of my existence. As much as I try to savour the freshness of each initial event, I also try to be mindful and seek out the wonder in the tapestry of the familiar.

I continued my journey west from South Luangwa National Park towards Livingstone, a town where I had lived almost sixteen years earlier. I wondered how well my recollections would fit with the reality I was about to encounter. Memories, once vivid, can become hazy and distorted as the years slip by.

I thought back on all the places I had visited earlier on this journey while still travelling with Sarah, places we had planned to revisit one day together. She would always be an integral part of those experiences, which were still crisp and clear in my mind. How would these landscapes of memory transform in the absence of her companionship?

My future now awaited a solitary traveller.

Unable to complete the journey of over 1,200 kilometres in less than one day, I took the opportunity to break the long ride into two parts, stopping briefly in Zambia's capital city, Lusaka, arriving as night fell. I followed my rudimentary map to the Wanderers Overland camp, pleasantly surprised to find the streets well-lit and signposted. Pulling into camp, I received a warm welcome from a fellow rider who had just emerged from a gruelling 2,000-kilometre traverse of the Democratic Republic of Congo. His bike was in pieces as he rebuilt its engine before embarking on the next leg of his journey. As often happened with my fellow travellers, I felt an instant bond form between us over our shared experiences and our common style of travelling.

We talked late into the night about our adventures and mishaps.

The next day, as I explored the city, it became apparent that I had entered a modern, well-ordered capital, contrasting markedly with many of the cities I had seen along the way since entering Africa. I typically was not a fan of big cities, but the prospect of finding several motorcycle parts overdue for replacement kept me there for a few days. In the end, the search for parts proved fruitless, but it did provide a comfortable respite from the long ride across Zambia.

I made the final push towards Livingstone on a Sunday morning, heeding the warnings of locals who had advised me that many drivers on the road were often still drunk after their Saturday-night revelries. Thankfully, the roads were quiet, and by mid-afternoon, I was nearing my destination, scanning the horizon for the first sight that had greeted my arrival so many years before.

The town of Livingstone is close to Victoria Falls, an enormous kilometre-and-a-half-wide waterfall where the mighty Zambezi River cascades into the narrow Batoka Gorge, creating a plume of mist that can be seen from afar. The locals call the falls *Mosi-oa-Tunya,* which translates to "the smoke that thunders." It was explained to me that the column of spray that rises through the air resembles a cloud of smoke, and the noise of so much water, crashing against the rocks at the base of the falls, fills the air with an ominous rumble.

I barely recognised my old home.

Making several passes along the main street, I looked for familiar landmarks and could hardly find any. I had arranged to meet a former colleague at a local backpackers hostel, which I struggled to locate. Turns out, it no longer occupied the same building, having moved to bigger, more luxurious premises. Eventually, I was forced to stop and ask for directions.

Before long, I was catching up with long-lost friends and acquaintances from my days as a raft guide on the Zambezi River. As luck would have it, a few seats were available on a raft leaving the following morning for a quick trip through the first ten major rapids. I jumped at the chance to revisit the river. After a blurry night of countless bars and endless tales from the river, I awoke feeling rather sorry for myself. I donned a couple of extra thermal layers and joined Grubby's Extreme Rafting trip into the Batoka Gorge.

Whitewater rafting is one of the most effective hangover cures in existence.

After several wet and wild rapids, I began to feel human again. Little had changed since I had last ran the river, so many years before. The sheer power of the water was as impressive as ever as it squeezed its way between the towering, dark canyon walls, responding to every constriction with a tremendous fury of whitewater, confirming my long-held belief that this was still one of the best one-day rafting trips on the planet. As we successfully negotiated rapids with names like "Stairway to Heaven," "Devil's Toilet Bowl," and "Gnashing Jaws of Death," it became apparent that the talented guide was heavily supplementing my mediocre paddling efforts.

Rivers used for whitewater recreation throughout the world are classified from Class I to VI. The classification system was designed to help river users gauge the danger and difficulty of individual rapids, with Class I being the easiest and Class VI the most hazardous. Most of the rapids on the Zambezi are Class IV and V, with Class V generally considered the upper limit of navigability on commercial whitewater rafting trips.

By midday, we had reached the infamous Class VI rapid, "Commercial Suicide," a mandatory portage where rafts must be dragged across the polished rocks to avoid one of the biggest rapids on the river. We stopped on the river bank for lunch and gazed in awe at the overwhelming power of nature, its raw energy on savage display.

I shuddered when I thought back on my old life here.

While there, as a guide and safety boater, I kayaked that rapid almost every day without so much as a second thought. How easy it had been to get so comfortable with something extremely dangerous and still feel completely in control. I thought of Hendri Coetzee and his unexpected death on the water. I also thought of all the close calls I had had over the years; watching the water surge and swirl impressed upon me a timely reminder of the fragility of life.

On returning to Livingstone, I bumped into an old friend who owned another rafting company specialising in multi-day trips on a variety of rivers around the world. He had a four-day Zambezi trip leaving the following morning and asked me if I'd like to come along and help out. I jumped at the chance to run the river again, this time in a kayak. Without any of the essential gear needed, I spent the rest of that day digging through piles of old equipment discarded by visiting paddlers throughout the years. Eventually, I assembled the necessary kayak, paddle, sprayskirt, helmet, and splash top.

Before I knew it, I was back at the "Boiling Pot," the first rapid at the very base of Victoria Falls, thoroughly soaked by the thundering mist that arose from within the deep chasm. I'd had little choice in the types of kayaks on offer, and my selection – a small, edgy play-boat – was making me rather apprehensive as I compared it to the larger-volume boats of the other safety kayaker and the video boater.

The river and I danced together for the next four days, sharing the lead in a furious waltz of whitewater.

Sometimes the river would dominate, spinning me this way and that, but as I settled into its rhythm and recalled the steps required for each rapid, I began to take more control, gliding downstream in harmony with my surroundings, stumbling less often. We spent peaceful evenings camping on deserted beaches, sleeping under the vast African night sky to the sounds of the river, and enjoying fine food and good company around crackling campfires.

Early in the afternoon of the second day, as our flotilla of boats gently drifted through one of the calmer stretches of river, I paddled ahead of the group. Enjoying the serenity and solitude, I occasionally glanced over my shoulder to make sure I wasn't getting too far from the rafts. With no significant rapids in sight, a relaxed atmosphere settled upon me.

I allowed my mind to wander.

A large raptor circled high above the canyon, soaring on columns of warm, rising air. The current bubbled and murmured as it eased its way downstream between the steep canyon walls. I was lost in thought when an audible whoosh disturbed the water behind my kayak.

I assumed the rafts had caught up with me, and I looked upstream, expecting to see them. They were over half a kilometre behind me. Something large and muscular was churning through the water less than a metre from the tail of my kayak. I had already turned to face forwards and begun to paddle as my mind raced to process what I had just seen. It wasn't a raft.

It was a crocodile . . . coming straight for me.

I instinctively increased my pace before turning again to confirm what was happening. Without missing a paddle stroke, I looked over my shoulder into the eyes of something that appeared prehistoric. The monster stared back with an indifference that caused a chill to run down my spine. I was nothing more than a potential meal to this magnificent and terrifying beast.

"Croc! Crrrroooocc!" I screamed repeatedly at the top of my lungs as I shifted my efforts into overdrive.

I hoped to alert the boats behind me that we had some unwelcome company. It wasn't uncommon for a client to let a hand or foot slip into the water on the calmer parts of the river when the midday heat became oppressive. I doubted they could hear me, but I had to try. I thrashed the water to a froth, giving each paddle stroke every ounce of energy I possessed. Adrenaline coursed through my veins, and the thump of my heartbeat pounding in my ears drowned out all other sounds. I stole another glimpse behind me, convinced that nothing could possibly rival my frenzied pace. Yet there it was, exactly the same distance behind my kayak, easily matching my speed. Hungry, green eyes met mine, and I felt an unexplainable connection.

If it truly wanted me, it could have taken me.

A large crocodile can comfortably outperform a kayak, and this one appeared to be over four metres long. I reached deeper inside myself and tapped into the very last of my energy reserves, hoping I had the stamina to maintain this gruelling pace. By that point, I felt like I was paddling through treacle, and every muscle I used began to throb and burn. I knew I couldn't give up, but my strength was wavering. I wondered how long the chase had been going on. *Was it seconds . . . minutes?* It felt like hours.

Suddenly, a rapid appeared ahead, and a glimmer of hope emerged. A distant memory fought its way into my frantic mind, something about crocodiles not liking moving water. As I entered the first gentle waves of the imminent rapid, I risked another look behind. It was true. The crocodile had abandoned the chase.

I was safe, for now.

Utterly spent, I turned to face the turbulence, hoping it would be a simple rapid. After successfully navigating through the channel, I pulled into a calm pool below and waited for the rest of the group to arrive. When they saw my face, they immediately guessed what had happened. I suspected my complexion held little colour.

We were able to joke about it around the campfire that evening, but later, as I rested my weary body on a sleeping pad beneath the stars, I relived the experience over and over. Each time, my thoughts would return to that brief moment when I looked into the eyes of a predator, and I was its prey.

All too soon, our four-day adventure was over.

We reached our take-out point, and the thumping sound of a helicopter, our ride back to civilization, broke the serenity we had settled into on the multi-day trip. We loaded ourselves into the sturdy belly of an iconic "Huey" chopper with its doors latched open. The pilot flew us back upriver, skimming the water's surface, banking hard left and right through the twisting gorge before swooping up and over the majestic falls where our journey had begun.

Exhausted and exhilarated, I was full of regret about having to move on so soon as I prepared to depart on the next leg of my journey south. There were still several old friends in the area with whom I had wanted to reconnect. But with other commitments elsewhere and a bike in desperate need of a good service, I was determined to reach South Africa, where parts were said to be readily available. Somewhere on the road between Livingstone and the border crossing at Kazungula, my bike hit its 50,000th mile. I purchased it in California with just over 10,000 miles on its US odometer. The signs of excessive use in tough conditions became more evident each day on the road. I was approaching the border as I contemplated this significant milestone, and I let my attention wander, causing me to make another dumb mistake.

The crossing was smooth from Zambia to Botswana on a small overloaded ferry across the Zambezi River until we disembarked. Then, I failed to stop at the mandatory Ebola screening checkpoint. Crossing borders had become such a routine experience that I tended to ignore the swarms of people flagging me down. More often than not, they were offering help with completing relatively simple paperwork while charging an exorbitant amount for their assistance.

On that occasion, I assumed the medical examiners were part of a similar operation. I guess their crisp white uniforms should have given me some indication as to their intentions. Borders were often a little chaotic, and my mind was routinely preoccupied with how best to approach the customs and immigration officials. When the medical team finally caught up with me, I was patiently awaiting my entry stamp inside the immigration office. They instructed me to return to a small tent near the ferry. Again, in my full riding outfit and with a good distance to cross beneath the hot sun in search of this tent, I figured my temperature would likely be well above average by the time I got there.

"Why did you not stop at the Ebola screening?" asked the nurse. "This is very serious, you know. Now, I must check your temperature."

I shrugged as she held up a modern infrared thermometer and pointed it at my forehead. As she read the result, her eyes flickered between mine and the screen – just like last time. She tapped the side of the device with a painted fingernail and took a second reading, then a third.

"Ah!" she exclaimed. She turned to show the rest of the medical team my temperature.

That raised concerns among the small group of doctors and health officials. The four of them looked at me and began conversing loudly in a language I couldn't decipher – except for one recurring word: "Ebola."

"So, which of these countries have you visited?" said an older, well-dressed gentleman, pointing to a faded map of Africa taped to a dog-eared piece of cardboard.

"Here, here, here, and here, and there . . . and there. Oh, and there, too." I trailed my finger along the route I had ridden.

They looked uncomfortable as they bombarded me with a list of questions about my physical condition and travel history. Each answer was met with a gasp since I had ticked many of the boxes qualifying me as a potential carrier. A feverish discussion took place while I sweltered in the midday heat.

Finally, the more senior-looking member of the party looked me in the eye and, in perfect English, asked, "Do you have Ebola?"

"No," I answered sincerely.

They all breathed a sigh of relief and, in chorus, said, "Welcome to Botswana."

"At least, I don't *think* so," I whispered to myself as I reclaimed my place in the line at the immigration office.

My plan was to make it further into Botswana on that first day, but the allure of the nearby Chobe National Park was too much to resist. I turned off the main highway, and by mid-afternoon, I had found a comfortable camp in the small town of Kasane. I fought the temptation to cool off in the nearby river – signs warning of crocodiles and hippos were enough to discourage me. Feeling a rumble in my stomach, I decided to go for a walk through town in search of food instead.

As I wandered along the main street of Kasane, lost in my thoughts, I caught sight of the distinctive shape of a fully-loaded adventure motorcycle parked outside a grocery store. Hard panniers, similar to my own, displayed a map of the world, and an Australian licence plate betrayed the bike's origin. I scanned the street for the owner, excited to meet a fellow traveller and share information on the roads ahead.

Before long, a young couple emerged from the store wearing bulky overland riding gear, arms loaded with fresh supplies of food and drink. I introduced myself as they packed their bike with a well-practised efficiency that suggested they had been on the road for at least a couple of months.

Tanya and Dean were travelling north, beginning a three-year-long adventure that would take them to Europe, Asia, and the Americas. We traded stories, contact details, and some local currency as they were en route to the border. They were both giddy with excitement about what lay ahead, and their contagious enthusiasm reminded me of how I had felt at the beginning of my journey. As they completed their preparations for departure and donned their helmets, we shook hands and bade each other farewell.

Dean looked me in the eye, patted my shoulder, and said, "One life is all you get, right?"

I thoughtlessly responded with, "First pub on the right, mate."

Those two Australians reminded me of my adventurous friends from that earlier chapter in my life, the ones I'd send off with that saying in case I never saw them again.

After watching them depart and disappear into a cloud of fine dust, a growl from my stomach reminded me why I had walked into town. I always chose to eat where the locals did, and several people had recommended a small diner named Martha's Kitchen. I arrived there at four in the afternoon, the sign outside indicating it would be open until seven that evening. When I asked for a plate of the local stew, they told me they had already run out. When I asked for another menu option, they said they were also all out of that. When I jokingly asked if they had *any* other food, they told me they had nothing left to offer. Everything had been eaten. I asked them what they planned to do until 7 p.m.

They replied, "Wait until closing, of course."

We talked for a while and mused about my predicament as my stomach grumbled. Eventually, they suggested I try another nearby diner, but I had an almost identical experience when I arrived there. It was a rather stark contrast to how things function in the West. In Africa, they waste nothing. When food ran out, they would not prepare more for fear that it may not be eaten, whereas we would throw good food away rather than disappoint a customer. It reminded me of an old Irish expression I often heard growing up: "Wanton waste leads to wasteful want." That principle was applied throughout the continent, and the people were most industrious at recycling and repurposing everything, primarily through necessity.

***

Weeks later, I began to receive frantic enquiries via email about my well-being from family and friends. The news was slowly filtering out of Africa about a motorcycle accident involving a couple of overland riders. With a deep sense of sadness, I gradually discovered the truth. While riding through Uganda, Tanya and Dean were fatally injured when a truck carelessly crossed into their lane and collided with their bike.

In that fleeting moment of conversation, we were oblivious to the twists of fate that awaited us. We knew what we were doing came with a degree of inherent risk, but we understood that the very foundation of adventure relies on the outcome being uncertain. We accepted that – even embraced it. After all, what worth is an expedition if every step is predetermined and guaranteed? We understood that our spirits truly come alive in the face of adversity and unpredictability. Only in the depths of uncertainty do we discover our resilience, the vastness of our potential, and the true measure of our existence. Within the realm of the unknown lies the very essence of what it means to be truly alive. Tanya and Dean had chosen to dive headfirst into the unknown, to forgo the security of a settled life, and to taste all this world had to offer.

Dean was right – one life is all we get. It's essential to make every second matter.

***

From the Chobe region of northern Botswana, I continued south along the lonely highway to Francistown. Skirting the edge of the park, I occasionally saw large herds of elephants purposefully walking across the veldt, like a flotilla of ships sailing across a sea of tall grass. Warthogs and baboons scurried across the road ahead while spiralling raptors soared effortlessly upon hot, thermal updrafts. It was a quiet road with little other traffic and very few potholes, so I could allow my mind to wander, a welcome respite from the hours of intense concentration required on most of the routes I had ridden until then.

I took the opportunity to refuel and take a break at the tiny village of Nata, where a group of travelling seed salesmen took time to explain why the average I.Q. in Ireland had been steadily dropping over the years. According to them, the Catholic church had been selecting only the finest minds to serve as priests and nuns, which had a quantifiable impact on the general population by removing Ireland's smartest from the gene pool. It was an interesting observation, one that I was completely unaware of.

"Probably," I assured them, "because my I.Q. is so low."

It had been a long day's ride, and as I pulled into a camp north of Francistown, Botswana's second-largest city, I barely had time to erect my tent before witnessing yet another stunning African sunset, the blood-red sun setting the horizon on fire in a blaze of colour. Life in a tent is dictated by the sun, and by 6 a.m. the following morning, I was wide awake and breaking down my camp.

The daily pre-ride inspection of my bike revealed a very slack chain and an over-worn rear sprocket. I tightened the chain to its last adjustment and hoped it would get me as far as South Africa where I had a new one waiting for me. However, several kilometres of bumpy, dirt roads dashed my hopes when my chain popped off on more than one occasion. Each time I hit a large pothole, the rear shock absorber would compress, throwing the chain from the sprocket, causing a sudden loss of power, followed by a loud grinding sound of metal on metal. I had to engage the clutch quickly to avoid causing serious damage. I knew I wouldn't get far with my bike in this condition.

I had simply pushed it too hard.

I nursed the machine slowly into Francistown and set about searching for a garage where I could find the specialised tools required to shorten the chain by removing chain links. It didn't take long to locate a few fellow bikers, and after a couple of phone calls, I was introduced to Joe de Souza, a mechanic who kindly allowed me access to his workshop.

With the right tools, I had the old chain off and shortened and was soon back on the road. After stopping to refuel, as I pulled out of the service station, my main fuel line ruptured, spilling flammable petrol all over the hot engine. With Joe's help, I located an adequate length of fuel line and soon had that problem fixed, although it left me wondering what else could possibly go wrong before I got to a country where I could find plentiful spare parts. It was getting late in the day and I was tired from being bounced around by the rough roads and constantly fixing broken things.

I decided to make a run for the border despite the time.

Crossing into South Africa from Botswana through the Martin's Drift border post was relatively effortless. However, darkness had already descended by the time I completed the customs and immigration formalities. I found a quiet spot by the edge of the road where I could sit on the curb and allow the significance of this crossing to settle in, knowing this could be my last border on my epic voyage. My plans to enter Lesotho still hung in the balance, dependent on the accuracy of persistent rumours about a military coup in the region. Ultimately, my destination lay in Cape Town, Africa's southernmost city.

As I sat in the warm night air, a gentle chorus of crickets brought a tranquil calmness to my thoughts. I contemplated the remarkable journey that had brought me there. Sixteen months on the road, and the end was tantalisingly close. Temptation whispered for me to keep moving but the hard-earned wisdom from countless African roads cautioned against travelling after nightfall with the perils lurking in the shadows. The combination of wild animals and treacherous potholes made it a dangerous gamble. So just a few kilometres from the border, guided by a mixture of prudence and reluctance, I sought out a quiet campsite and settled in for a fitful night's sleep.

Beneath a canopy of stars, a surreal blend of exhaustion and anticipation enveloped me. Dreams melded seamlessly with reality as I surrendered to the night's rhythm. The stillness was punctuated by the calls of distant wildlife and the soft breeze whispering through the trees. I had entered the final chapter of an extraordinary odyssey.

CHAPTER THIRTY-FOUR

The Kingdom in the Sky

Personal Diary Entry: 1st October 2014

Day 506 (of 549)

58,479 km of 62,840 km (36,337 mi of 39,047 mi)

The final frontier . . . this journey has tested my limits and shaped my soul. As I sense the curtain slowly descending on this chapter of my life, I can hear the hushed overture of a new beginning. Just as every sunset casts the shadows that herald the sunrise, so must I prepare for a new dawn. Soon, I will turn the page and begin afresh.

Waking early to the sound of birds greeting the new day is one of the simple pleasures of life in a tent. As the sun slowly rose and the world filled with light, I quietly rested longer than usual, letting the realisation of where I was fully sink in. Just south of the Limpopo River, I was inside South Africa, and Cape Town, my goal, was within reach. As I packed my tent and loaded my bike, I prepared for the short drive to the nearby city of Johannesburg, where I intended to see a friend.

Travelling south, I tried to ignore the grinding that had grown significantly as my chain and sprockets far exceeded their intended lifespan. I knew I could find replacement parts soon, and I hoped the ones I had would hold out for the last few hundred kilometres. A broken chain at high speed can cause catastrophic damage as it whips into the engine casing. Taking it easy when entering the fast-flowing traffic of South Africa's motorway system – so that I didn't become a hazard to myself and those around me – made progress difficult.

By early afternoon, I had reached the residence of a dear friend I had worked alongside many years before in Chile. For the next week, Lynn and her delightful mother opened their home to me, showering me with overwhelming hospitality as I set about finding the parts I needed to put my bike back into reasonable running order.

Finding spare parts in Africa had been a constant headache, and it felt odd to have so many choices when locating what I needed in Johannesburg. The last time I'd visited any kind of motorcycle dealership was in Egypt, over 8,000 kilometres away, and somehow, I had made the parts I was carrying last this long. A few simple phone calls to nearby suppliers left me with an extensive list of options.

I began with the BMW stores, but, as always, their prices caused me to reconsider, confirming my belief that BMW was an acronym for "Break My Wallet." The bearings I required, seven in total, could all be found at a local specialist store for the same price that BMW wanted for one. All in all, I found everything I wanted for a third of the price they had initially quoted.

I then spent several days in Lynn's garage, dismantling my bike and slowly rebuilding it, before taking it for several test rides. After replacing all the worn parts, it felt like a new machine and became a pleasure to ride again. I'd almost forgotten how smooth it could feel to have everything working the way it should. The continent had been hard on the bike, but it had still outperformed all of my expectations, handling the tough conditions with an ease only limited by my questionable abilities.

But it wasn't all work and no play while in the city.

Lynn introduced me to several charismatic members of the Exploration Society of Southern Africa (ESSA). With them, I had an opportunity to kayak on one of the local rivers just outside the city. Even though the water was low and it wasn't the cleanest river I'd ever been on, it felt great to be back in a boat. Floating down a river in a kayak has always struck me as one of the most unique ways to experience a region, and that was no exception.

The days flew by, and I soon realised that the thirty-day visa I'd obtained at the border on entering wouldn't be long enough. Lynn's extensive knowledge of her home country had me thinking there would be much more to see than I had initially thought. When I first entered the country, it was late in the day after a tiring ride, and the immigration officer had asked me how long I planned to stay. I'd asked for the maximum time allowed, and he stamped my passport with a visa valid for one month. Sometime later, a quick internet search revealed that Irish citizens were entitled to a maximum of ninety days.

At the time, I had no idea.

That was the thirty-eighth country I had entered on this voyage, and I certainly did not know the visa regulations for each one. If an immigration officer told me that the maximum stay was thirty days, I tended to believe them and plan accordingly. While in Johannesburg, I made a quick visit to the local immigration department, located in the heart of the notorious Central Business District. The two senior officers on duty informed me that the only way to correct the mistake made by the border guard would be to revisit the place I had crossed and exit the country.

I'd always had a strong desire to visit Lesotho, but as I got closer and heard the reports of trouble in the tiny country, I had revised my plans. With the refusal of the immigration officers to offer a more practical solution, I resolved to alter my route to include the "Kingdom in the Sky." That would allow me to exit the border and get restamped without returning to where I came from.

With all of the maintenance to the bike completed, I decided it was time to move on, but a last-minute phone call from a member of the ESSA group caused me to reconsider. One of their members had suffered an injury just before departing on a five-day hike through the Hluhluwe-iMfolozi wildlife reserve, and he was willing to sell me his place on the hike for half the price. Opportunities like that did not come along often, so I was always glad to retain the flexibility to seize them when they did arrive. I jumped at the chance to join the hike and set out for the east coast the following day.

The ride to the coast took me through the Mpumalanga region and into KwaZulu-Natal. Avoiding the main highways, I kept to the twisting back roads through rolling hills and fertile, sugarcane farmland. A thick haze filled the air as farmers burnt the last of the winter's growth in preparation for the following year's planting. It added an eerie edge to the ghostly atmosphere as I passed by long-forgotten battlefields from the Anglo-Boer wars.

The ride took longer than expected, and it was well after dark by the time I reached my destination. I spent a few days on the coast in the little town of St. Lucia, which lay within a unique world-heritage site that contained five ecosystems and was home to over ninety percent of South Africa's natural crocodile population. At night, hippos roamed the streets – I almost rear-ended one on the motorcycle – and a healthy population of sharks patrolled the estuary.

I decided to swim anywhere other than here.

Over the next few days, seven intrepid members of the ESSA group arrived, and we set off into the nature reserve for our five-day "primitive" trek. Carrying everything we needed in our backpacks, we were joined by our local Zulu rangers, Nunu and Nantabela, before entering the untamed bush. Hluhluwe-iMfolozi is Africa's oldest wildlife reserve, established in 1895 to protect the Southern Black Rhino from extinction; its dedicated and well-armed guardians had enabled its residents to survive, even though the constant threat of poaching remained a major concern.

Leaving behind our cell phones and watches, we slipped into a peaceful rhythm of rising and sleeping with the sun and moon, quietly walking through the park so as not to disturb the local wildlife. Each day, we would follow our guides over the rough terrain, relying on their keen senses to help us spot the animals all around us. By noon, while the sun was at its hottest, we would find shade, eat a simple lunch, and rest until the air cooled. Then we resumed our hike before finding a suitable camp for the night, always near a river bed. Digging several feet into the cool, damp sand would expose the earth-filtered groundwater, clean enough to drink untreated.

As darkness fell, one of the guides started a fire, and then they began preparing a basic dinner as we rolled out our sleeping mats to rest beneath the stars. It was such a simple pleasure to chat quietly while enjoying a nourishing meal, sharing stories from the many diverse journeys that had brought each of us to this moment in time and space. Throughout the night, we took turns to stand watch and keep the fire burning to discourage any nocturnal visitors from coming too close. On the third night, halfway through my watch, two male lions came within a hundred metres of our camp. One of them roared to announce his presence.

The lion's roar awakened something deep and primal within all of us.

Everyone in camp suddenly lost all interest in resting and peeked out from their sleeping bags. Several tense minutes went by before the lions finally moved on. The reserve was home to the Big Five: elephant, buffalo, lion, rhino, and leopard. Of these, only the cunning leopard remained elusive – the others made several appearances on numerous occasions throughout our five days.

Returning to civilization afterwards took some adjustment. However, the serenity that had settled upon us while out in the bush was still with me. I'd left my bike outside the park and was relieved to find it still standing when I returned. The park staff had warned me that elephants liked to wander around the complex at nighttime, getting up to all kinds of mischief. They could have taken personal offence at my bike being there and chosen to knock it over or trample it. I packed my belongings onto the bike and took to the road, riding roughly in the direction of the infamous Sani Pass that leads from South Africa into Lesotho.

When I shared my plans to enter Lesotho with a fellow traveller, I received warnings not to underestimate the complexity of entering via the route I had chosen. The Sani Pass climbs steeply through the jagged Drakensberg Mountains, reaching a height of almost 3,000 metres before piercing the border of the mountain kingdom.

The warnings only served to intensify my determination.

I found a quiet backpackers hostel at the foot of the pass to spend the night there as poor weather had shrouded the sinister-looking mountains in ominous clouds. If I was going to attempt this route, at least I wanted to enjoy the views.

Finally, the weather cleared, and it was time to leave. With bright sunshine breaking through the clouds, I put on my warmest gear and pointed my bike uphill. At the base of the pass, I exited through the South African border post, where they questioned whether or not my bike would make it to the top, leaving me feeling more apprehensive than I already was. I wondered if it really was as bad as they made it out to be.

The dirt road got gradually steeper, but the new rear tyre I had fitted while in Johannesburg handled the loose gravel with ease. Each hairpin bend provided me with a spectacular panorama of the valley below. An hour after starting, I was at the top, feeling euphoric even as strong, bitterly cold winds whipped over the summit.

I found myself in a strange land, a unique island of mountainous plateaus nestled in the middle of South Africa, a beautiful anomaly amid a region of modernity. Locals lived in artfully constructed round stone huts with thatched roofs; they travelled on horseback wrapped in thick, brightly coloured woollen blankets. In the east of the country, the turmoil rumoured to be taking place in the distant capital was virtually unheard of among the people I spoke with. I spent several days driving westwards through the sparsely populated countryside, wild-camping alongside pristine mountain streams. At these higher elevations, the early mornings were crisp and frosty. Emerging from the warm cocoon of my sleeping bag took longer than usual.

As I approached the capital city of Maseru, I began to hear whispers of "popular uprisings" and "military coups." These appeared to be groundless rumours. Entering the capital, I found little evidence that the simmering friction was anything more than political manoeuvring by the elites. Other than a few plumes of dark smoke rising over the city skyline, the region seemed peaceful; people were going about their business as though it was just another day. The current prime minister appeared to be involved in a minor power struggle with the former head of the military and the old king. Neither side would make the necessary compromises to break the impasse to the detriment of a country they all claimed to love so much. Watching the residents of this modest city go on with their lives while their leaders squabbled left me pondering the futility of our social hierarchy.

Why do those in charge often place their needs above those who chose them to lead?

As I neared the end of my journey and reflected on how it all began, including the dark moments that had tested me along the way, I sometimes needed to remind myself that, as challenging as my journey had been, it paled into insignificance when I looked around and saw the struggles faced by some who inhabited this vast continent.

CHAPTER THIRTY-FIVE

A Growing Sense of Reluctance

Personal Diary Entry: 15th October 2014

Day 520 (of 549)

60,756 km of 62,840 km (37,752 mi of 39,047 mi)

There are times when I find the contemplation of doing something to be more difficult than actually doing it. I have mornings when I stare at my maps, pondering the ride ahead and the complications of facing the unknown, but once I am on the bike and moving forwards, everything seems to simply fall into place.

With the end in sight, a whirlwind of emotions stirred within me. Like a coin tossed in the air, there were two sides to how I felt. At times, I wanted my odyssey to last forever, but there were also days when I was ready to see it all end. The delicate balance between the longing for perpetuity and the acceptance of impermanence imbued my journey with a profound sense of depth and meaning. Occasionally, the weight of the road, the distances traversed, and the uncertainties that accompanied each new dawn wove threads of fatigue into the fabric of my consciousness. In those moments, a yearning for the familiar comforts of home, the tender embrace of a loved one, and the tranquillity of a settled routine would flicker through my mind.

After almost a week in Lesotho, I returned to South Africa via the Maseru Bridge crossing near Bloemfontein. While on the road, some friends made contact and invited me on a four-day canoe trip down the Orange River, which borders Namibia, so I had a firm destination and a date to be there. I wouldn't pass up another chance to be out on the water.

Leaving Lesotho was effortless. The border guard gave a cursory glance at my passport and waved me off. Entering South Africa for the second time was a bit more difficult. The thirty-day visa I'd been mistakenly granted on my first entry was about to expire, so the immigration officer was unwilling to let me return.

Of course, I had expected there to be a problem.

After I visited the immigration headquarters in Johannesburg, where none of the officials were willing to give me a definitive answer on how to resolve the initial issue, I knew there would be at least one bump in the road. In preparation for this, I had arrived well-fed and hydrated, with a positive attitude and a big smile. I'd learned a thing or two from all the runarounds I'd experienced.

Thankfully, the guard on duty agreed when I asked her to consult her superiors. She disappeared with my passport while a long line of patient travellers began to form behind me. After thirty minutes, she reappeared and promptly stamped my passport with an additional sixty-day visa. With the large crowd that had gathered behind me, I breathed a sigh of relief and was on my way again.

After the rougher roads of Lesotho, it felt great to open up the throttle on my bike once more and speed across the flat, boundless plains of the Free State. By lunchtime, I had arrived at South Africa's judicial capital, Bloemfontein, where I took a brief break to eat and refuel. With plenty of daylight left, I decided to push on towards the Northern Cape province, stopping for the night in the small city of Kimberley.

In 1871, prospectors discovered diamonds on the slopes of the small hill where Kimberley stands today. News of the discovery attracted thousands of miners and prospectors to the region, and soon, the hill became a hole as frantic digging ensued. Kimberley became the site of "The Big Hole," claimed to be the largest hand-dug hole in the world at a depth of 240 metres. Almost 3,000 kilograms of diamonds were extracted from this mine, consolidating the fortunes of men like Cecil Rhodes and the De Beers brothers.[28] The De Beers Corporation retains a monopoly over the world's diamond market to this day.[29]

South Africa had changed since I travelled there sixteen years before.

The noble ideals upon which the African National Congress swept to power had been gradually replaced with a kleptocracy, fattening the few at the expense of the many. Money intended to support and improve the country's infrastructure disappeared into the pockets of thieves and hypocrites until the pot ran dry, and upturned hands were again presented before the unscrupulous World Bank, the International Monetary Fund, and predatory foreign investors. The brave cadre of freedom fighters – or terrorists, depending on which side you supported – were strangely quiet, and the current leadership maintained some questionable beliefs.

The current president, while defending a rape allegation from an HIV-positive victim, claimed that showering afterwards would prevent him from catching the virus.[30] In a country wracked by AIDS, a senior health minister advised the general public that a diet of garlic, beetroot, lemon, and potatoes could help prevent the disease.[31] With leadership like that at the helm, I could only wonder what course the country would take. For the chosen few, conditions in the country had improved beyond their wildest dreams, but for the majority, little had changed.

***

From Kimberly, I continued to ride west into the barren Kalahari Desert, a sparsely populated region of acacia trees, dry savannah, and rocky red dunes. I'd arranged to reconnect with my friend, Lynn, in the small town of Upington, close to the Namibian border, before embarking on a four-day canoe trip down the Orange River with her and the Warriors Program. The organisation took students on their gap year and introduced them to a world of adventure activities while building confidence, fitness, and environmental awareness.

After consulting guidebooks and local rafting companies, the chief facilitator felt that a couple of extra hands would be useful should the whitewater prove too rough for the group involved. In the end, the river was well within the capabilities of all those involved. It was a pleasant voyage as we floated through the spectacular Orange River valley, completely removed from civilization, accompanied only by the sound of our paddles slicing through the river's surface and the songs of a seemingly endless variety of birdlife. Each night, we would find a suitable beach on the Namibian side of the river and take some time to explore before camping under the stars.

After four wonderfully peaceful days on the river in the company of some inspiring, young individuals, I returned to my bike. I resolved to push south towards the coast through the Great Karoo, a semi-desert area occupying a vast swathe of the central region of South Africa. Its harsh climate had left it virtually uninhabited.

Bitterly cold nights followed hot days, and I was glad to have my warm sleeping bag at the end of each day's ride. Huge distances separated small townships as I kept to the minor dirt roads crisscrossing the parched territory. Telegraph poles were often topped with the large, thatched nests of the sociable weaver bird, as lonely, skeletal windmills attempted to suck moisture from beneath the earth, their vanes spinning wildly in the relentless, dry winds.

Parts of the Karoo had been tamed.

Hardy shepherds tended their flocks, scraping what nutrition they could from a land that offered little. Karoo sheep, well-adapted to the severe conditions, grazed on the sparse but nutrient-rich plants. Dominated by low, shrubby vegetation, robust succulents such as the distinctive aloes dotted the landscape, storing precious water in their leaves to survive prolonged dry spells.

On my third day, I underestimated the distance to my final destination, and as night fell, I was deep inside the Great Karoo's southern mountains. The endless ridge lines turned deeper shades of orange, pink, and purple as the sun set behind me. With little idea of how much further I had to go, I decided to push through, hoping to reach a town where I could find a warm bed and a hot meal.

As the darkness consumed me, I rode into the narrow tunnel of dim light created by my filthy headlight, floating the bike atop the loose sand and gravel, maintaining a high enough speed where the undulations of the rough road seemed to even out for a smoother ride. Frequent patches of deep sand would occasionally pull my bike off course, but I resisted overreacting, gently twisting the throttle to power my way through, hoping to find solid purchase on firmer ground.

It could have been due to a growing sense of fatigue from relying on my wits and instincts for so long that, on this endless evening, I began to feel deeply disappointed in myself. Doubts invaded my thoughts, and I second-guessed every decision I made. Troubling anxiety nudged its way into my mind. I was taking unnecessary risks, and the possibility of having an accident while so close to the end frightened me. My route carried me south through a vast expanse of empty land, providing little protection from the strong crosswinds that buffeted my bike.

Likewise, I was beating myself up inside.

Somehow, my brain turned a regular day of riding into a "damned if you do, damned if you don't" scenario. I felt like it was a bad idea if I stopped to camp in this flat, barren land, with my body too tired to go through the motions of pitching a tent or cooking a meal. *Shouldn't I stop being so pathetic and just power through? What if I'm just a half hour away from a suitable place to sleep?* I also felt like it was idiotic to continue onwards when I should quit while I was ahead and get some rest. *What if something bad happens and I could've avoided it by stopping when I knew it would be smart to do so?*

Eventually, when I began to see signs for the next town coming up, I felt relieved and pleased that I had continued on, even if it was dangerous in the dark. I spent the rest of the drive fascinated by how quickly my mind had gone to war with itself over such a simple decision. I wondered what was going on beneath the surface. When I felt around and did a little digging, I discovered a painful rawness within me;

something I hadn't noticed for a while because I'd been busy distracting myself with friends and adventures.

There was something still haunting me. I decided to confront it later.

Battered and bruised, more inside than out, I found refuge from my thoughts and the savage winds at a comfortable guest house in the quiet town of Beaufort West.

The following day, as I approached the Cape Fold Mountains that separate the dry inland plateau from the ocean, I entered a series of twisting roads that carried me up and over steep ridges and into the fertile coastal region. I traversed through the spectacular Prince Alfred Pass, ignoring signs advising me to turn back because of flooding. I presumed I could ride through any washouts or, at worst, turn around. Even though the road had been badly damaged, I was able to squeeze through to the coast. Blue skies and turquoise waters revealed themselves as I descended out of the clouds.

The small resort town of Plettenberg Bay proved to be a comfortable stopover, enabling me to catch up with an old friend who lived in the area. Braden, a former housemate from my time in New Zealand, had moved to South Africa to start a family and manage the nearby Bloukrans bungy site. It felt good to see his familiar face again. After a couple of days in Plettenberg Bay, I continued west, and for the first time on my journey, I began to see signs for Cape Town.

My final destination was getting closer.

The coastal road wound its way through luxuriant forests alongside beautifully barren beaches, but I wasn't able to fully enjoy what I was experiencing. Something felt wrong. *Why isn't my heart swelling when I think about the impending grand finale?* The feelings just weren't there. In such a mood, I was extremely hesitant to cover the last several hundred kilometres that would complete my trip. I decided to pull into the quiet town of Mossel Bay to break up the last leg of the ride.

Mossel Bay is a sleepy harbour town on a stunning section of the Garden Route. It was home to many artisans and surfers, where the numerous beaches provided endless opportunities for riding the powerful waves that pummelled the coast. It was also the place where the first Europeans landed on South African soil in 1488.

Bartolomeu Dias and his Portuguese crew stopped briefly there while establishing a trading route to India to refill their water supplies from a freshwater spring. Before the sailors could make themselves comfortable, a surprise attack ensued, and a band of natives repelled them under a hail of stones and rocks.[32]

But what the town is famous for is rather unique.

In the centre of the old town, there stands a gnarly, twisted milkwood tree where mariners would deposit their mail inside an old boot in the hope that a ship passing in the opposite direction would collect it and carry it home. Known simply as the Post Office Tree, it was still in use, with a boot-shaped post box installed under its low-hanging branches. I spent a reflective morning in the shade beneath the old tree, writing the last of the postcards I would send to my father and nephew before slipping them into the old boot.

From Mossel Bay, it was less than 400 kilometres to Cape Town. If the roads were in good condition, which they invariably were in South Africa, I could ride that distance in just a few hours. But I procrastinated. I settled into a comfortable little hotel room and unpacked my luggage. I also began to unpack my messy feelings.

I had a growing sense of reluctance to see the journey's end.

Each day, as I neared the destination that was so close after being so far away, I felt overwhelmed with foreboding, anxious that I would stumble just before the finish line. I had yet to discover the source of this uneasiness. I suspected it could be related to my unwillingness to relinquish the freedom I had acquired since my odyssey began. I knew it would soon be time for me to pack my gear for the final time and return to work and a "normal life."

I planned to sell my bike and gear to raise funds to begin the next chapter in my life, and I knew it would not be easy to say goodbye to a motorcycle on which I'd had so many incredible adventures. I was surprised because I'd always thought I couldn't become attached to material things. I hoped I could find someone who wanted to take Uhuru on its next adventure. But there was more than that. More was troubling me than the thought of losing my precious bike or my coveted freedom.

It was Sarah.

It had been a while since I'd allowed myself to entertain any thoughts of her. But now, I stepped into them – cautiously. I explored my worries and let them form into questions. *Wouldn't it be a better ending if we were able to share it? What if I get there alone, and I feel like I'm lacking something?*

"Okay, good. Now, go deeper," I told myself. "What's this really about?"

I lay down my toothbrush on the sink ledge. I looked at myself in the mirror. I waited. I closed my eyes. *Ah, there it was.*

I feared that her absence at the end of such an important milestone in my life would break my heart all over again. I feared how my inability to forgive her, after all of these months, would turn into anger, when I reached my goal – what should've been *our* goal.

Now that I knew what the fear was about, all I had to do was face it tomorrow. I decided not to lose sleep over it. Perhaps, instead of heartbreaking, it would feel healing. Maybe that was how it was meant to be.

I could never have imagined how our trip would evolve into *my* trip. I often wondered what I would change if I could do it all over again, but life is too short for regrets. It is a delicate balance of embracing the present while learning from the past. The past, with all its triumphs and tribulations, could not be rewritten or undone. It was my teacher – imparting wisdom and shaping my understanding of this world. I could only reflect, try to understand, and grow through all the experiences that had led me to this moment.

Worrying about the future felt like a fruitless waste of energy. Challenges would arise, and I would meet them, give them my best, and hopefully triumph. If I didn't, I would look on them as opportunities to better equip myself for the next time.

I was reminded of a quote initially said in Portuguese by author Fernando Tavares Sabino but somehow attributed to John Lennon, Mark Twain, and Oscar Wilde: "In the end, everything will be okay. If it's not okay, it's not the end."[33]

CHAPTER THIRTY-SIX

A Quiet Catharsis

Personal Diary Entry: 13th November 2014

Day 549 (of 549)

62,840 km of 62,840 km (39,047 mi of 39,047 mi)

My year-and-a-half-long voyage from Cork, Ireland, to Cape Town, South Africa, has reached its conclusion. As hard as it was in my darkest moments, I had to keep reminding myself that time only moves forwards, solutions would present themselves, and there would always be a light at the end of the tunnel.

Riding from the quiet harbour town of Mossel Bay, following the tips I had received from local bikers, I hugged the coastline. The road clung precariously to the rugged cliffs, a winding black ribbon of asphalt snaking through the fragrant, verdant vegetation separating sea from shore. Before long, I began to encounter signs for "The Most Southerly Point in Africa, Cape Agulhas," and by early afternoon, I reached the Overberg, cruising into the tiny hamlet of Struisbaai. I set about looking for my intended destination. A young Greek couple I'd met many months before in Sudan had recommended a quiet backpackers hostel in the area, and when I pulled in, the owners immediately made me feel at home.

I had planned to spend only one night in this place, but the hospitality of my hosts and the wild, rugged seashore, reminiscent of Ireland's West Coast, cast a spell over me that was hard to break. Blustery winds carried a salty spray, soaking the white-washed walls of cottages sheltering under heavily thatched roofs. I decided to spend a little more time in the region, allowing myself to lose an entire morning sitting at Cape Agulhas, pondering the magnitude of that moment.

I could go no further south.

As I watched the waves crashing against the rocky shoreline, I enjoyed a tasty bottle of locally crafted stout. I was almost able to fool myself into believing I was back in Ireland at the very beginning of my journey. Beside me, an empty space echoed memories of times past, and I whispered to a ghost from my old life.

"We made it," I murmured and smiled. "I forgive you."

It was not at all like how I imagined it to be. There was no anger. There was no fear. No stirring in my heart at all – just peace. A quiet catharsis that, instead of undoing me, gently healed my wounds.

***

After a few peaceful days at Africa's most southerly point, I decided to continue west towards Cape Town, with an overnight stay in the busy surfing suburb of Muizenberg.

A few wrong turns took me into one of the region's many townships, a sprawling collection of temporary structures patched together with cardboard, tin, and plastic – a poignant reminder that South Africa still had a long way to go to redress the imbalance created by many years of social injustice and apartheid.

I felt like I had strayed into a part of this country I wasn't meant to see.

The poorly maintained road that cut through the heart of this ramshackle settlement provided a glimpse into the lives of its inhabitants. An older man with stubbly white hair, cloudy eyes, and weathered skin sat smiling and toothless on an upturned beer crate with his head cocked to one side as though in the process of identifying each of the thousand voices filling the air.

Hordes of dust-caked children in ragged clothes chased a homemade football through the narrow alleys that separated the shacks, screaming with delight each time they touched the ball. There were no apparent rules that I could decipher, just a chaotic pursuit interspersed with dramatic tumbles that would have seen a professional footballer writhing in agony. Without hesitation, these children would pick themselves up, ignore the fresh coat of sand clinging to their sweating bodies, and continue the game.

On several street corners, a *shebeen,* an improvised drinking den, kept many of the younger men occupied with cheap alcohol. Each shack pumped out its conflicting music preference, testing the limits of its sound system until all that remained was a distorted roar beneath a pounding rhythm. Even though it was still early, many of these young men were already drunk, stumbling onto the road ahead of me, glaring at my bike with unfixed gazes when I dared to use my horn.

And then there were the women, quietly gliding between streets and houses, often with the distinctive bulge of a newborn just visible beneath tightly wrapped colourful fabrics on their backs – always bearing a load, a sack, a bucket, or a package, either gripped firmly in strong hands or balanced perfectly on top of their heads. Through the open door of one roadside hut, I stole a glimpse of the world within – a heavy blackened pot bubbled over a charcoal fire set on an assiduously swept dirt floor, and fanned coals brightly glowed as fragrant steam escaped through the porous walls. From inside the dim interior, a young girl's sombre eyes locked with mine, seeming to peer into the depths of my being.

Who was this intruder staring back at her from the saddle of a motorcycle that cost more than she could ever hope to see in her lifetime? How dissimilar our lives were, our destinies decided at the moment of conception, our paths moulded by where we were born and the opportunities afforded us. Would I be any different from her had I been born into her life? I doubted it. Too often, I had listened to people accusing the poor of being responsible for their circumstances.

"If only they worked harder to pull themselves up by their bootstraps" – as though hard work alone could erase the disparities of fate.

We live in a time when the gap between rich and poor has never been greater, and the opportunities for bridging that gap have seldom been fewer. Is it any wonder some turn to desperate measures to cross that divide? When constantly bombarded with images of an abundant life that most will never realise through honest toil and perseverance, will shortcuts not be sought? As we witness those with obscene wealth protecting their spoils through nefarious means while destroying individuals, communities, countries, and our planet, is it not natural to question or challenge the status quo? Within the current model, for the few to have more, the many must have less. To that young girl, did I represent the obscenely rich? Was I part of the reason she had to make do with less?

I suddenly felt deeply ashamed of who I was and what I was doing.

Our global community felt broken, the once sacred social contract discarded. What have we allowed ourselves to become? It feels as though we have been fooled into living within a system that thrives on our collective discontent, perpetuating a cycle of longing and dissatisfaction. Each interaction with the modern world bombards us with reasons for unhappiness, subtly reinforcing the notion that true fulfilment comes from the acquisition of material things. Yet, deep down, our experiences teach us otherwise.

Breaking free from this pervasive mindset will be no easy task. At every turn, we face a legion of professionals, from software engineers to advertising executives and marketing experts, whose sole purpose is to shape our desires and manipulate our behaviour. They orchestrate a symphony of influence that coaxes us into conformity, while encouraging us to exchange our most precious resource – time – in pursuit of an elusive ecstasy, always just beyond our grasp.

Surely, a better world is possible.

As I rode from the township, I felt humbled and penitent for all the times I had complained or grumbled about my sorrows and struggles. I knew little of the real hardships an individual could endure.

316

***

Eventually, I reached Muizenberg and stayed briefly along the waterfront, lulled to sleep by the gentle hush of distant crashing waves. However, the stark contrast between this beautiful place and the township I had stumbled into tainted my appreciation for a destination catering to surfers and holidaymakers – people who had disposable income and time off. The cheerfully coloured beach huts and delightful cafés did little to lighten my mood or ease my discomfort.

I followed the coastal road to Cape Point National Park from the beachside resort, weaving my way south along the rocky peninsula. I spent a contemplative morning riding between the deserted white sandy beaches before hiking up to the redundant lighthouse atop the highest point. In perfect conditions, it can be seen for many kilometres, but an error in the choice of its location meant it was shrouded in mist for most of the year. Subsequently, a second lighthouse replaced it on a much lower point.

From Cape Point, I turned north for the last time, setting my sights on the city of Cape Town at the foot of the iconic Table Mountain. By early afternoon, I had reached my goal, but I couldn't resist riding a little further around the bay to a point known as Table View for a photo of the city dwarfed by its most recognisable landmark.

It was early evening when I resolved to find a place to stay for the night. With a tent, I usually didn't need to make reservations. However, each place I stopped at was either full or didn't offer camping, which proved to be quite serendipitous. When I finally did locate a place that could satisfy my simple requirements, I made some fortuitous encounters.

Another Irish biker, Maurice from Mullingar, had just arrived at the same time. He was about to embark on an almost identical journey to the one I had just completed but going in the opposite direction.

Later, while looking for a nearby motorcycle store, I had the good fortune of bumping into a local tour operator. He ran South African motorcycle expeditions through his company, Moto-Adventure, providing tailor-made itineraries for the intrepid traveller. After

317

opening up his home to me, he kept me entertained for the rest of the day, inviting me back the following evening to give an impromptu presentation to members of the local motorcycle club.

Through these contacts, I met many more local riders, and my schedule was soon full of invitations to rides and *braais* (barbecues) all around Cape Town. The hospitality I received from relative strangers was overwhelming and humbling and confirmed my belief that bikers are some of the nicest people on the planet.

Rides along some of the most stunning roads on the continent consumed my days. A morning spent with an old friend, Kate, who just happened to be passing through her hometown, saw us exploring the more fascinating parts of the city before hiking to the top of Table Mountain.

My plans to sell the bike seemed to be progressing well. I'd advertised it online and had been getting a lot of promising enquiries – until one afternoon, after a great ride over the Franschhoek Pass with new friends Rob and Hanlie, it refused to start. When I returned to my hostel in the city centre, I had parked it briefly in the street, and on trying to restart it to move it inside the premises, the engine would not turn over. The battery appeared flat, so after pushing it into the compound, I began exploring possible causes. I was hoping the alternator, which I had repaired and replaced previously, wasn't the cause of the problem. As it turned out, it was the alternator.

I had overcooked my copper doughnut yet again.

I put my plans for selling the bike on hold. I would never want someone to sell me a motorcycle that was not in good working order, and I certainly wasn't going to do that to someone else. It needed maintenance, and I was more than willing to spend more time tinkering with a bike that had served me so well for so long – but not right then.

Another adventitious encounter with a biker I had previously met in Kenya led me to a local shipping agent willing to crate my bike and send it on to my next destination. With time running out and my finances dwindling, I decided to hand the motorcycle over to Wolfgang at CD Shipping, hoping to see my trusty Uhuru again someday.

***

That moment I had been avoiding had finally arrived; the time had come to bid farewell to a land that had become my teacher, healer, sanctuary, and inspiration. Gratitude mingled with nostalgia, and a bittersweet sense of closure settled upon my heart. The people I had met there were living reminders of the immense power of human connection. They taught me that the truest wealth lies not in material possessions but in the richness of compassion and the willingness to share even when one's cup is nearly empty. Though my journey had reached its physical conclusion, I knew its echoes would reverberate with me for a lifetime.

As much as I had hoped to travel with an open mind, I carried prejudices and preconceived ideas about the people I met, which were challenged and reformed on every level into a more realistic impression of the world around me. I'd always noticed my tendency to judge others, to make assumptions based on fear or ignorance about what the people would be like in the next village, province, country, or continent.

On my Cork to Cape Town trip, I experienced the same thing as on my honeymoon motorcycle journey with Sarah through the Americas so many years before.

Despite similar warnings, experience after experience proved the stereotypes and stigmas wrong. Since then, I realised that wherever I go, our similarities far outweigh our differences. An unfortunate reality is that in the pursuit of power and influence, agendas are crafted that rely on a divisive narrative. Nationalism is often wielded as a weapon, intended to elevate one group while sowing seeds of fear and mistrust towards those who reside on the other side of an imaginary line on a map or those whose appearance or beliefs differ only slightly from our own. I have seen the beauty that emerges when borders fade and humanity takes centre stage.

What are borders anyway, but the festering wounds of imperialism?

In the laughter that transcends language barriers, the tears that flow from hearts burdened by sorrows, and the gestures of kindness extended without hesitation, I have witnessed the underlying unity and compassion that binds us all. Beneath the layers of culture, tradition, and belief, there beats a collective human heart that yearns for connection.

We are all of the same origins with the same basic needs. Beyond the physiological requirements for survival, we want to feel safe, loved, and valued; we want to feel like we are contributing to something greater than ourselves. Across the broad spectrum of individuals I had encountered along the way, I found these simple truths common among all of us.

The only trouble I've had on my journeys came from those in uniform – the police, the military, and the border officials. It seems as though donning those uniforms robs us of a little piece of our humanity. With authority comes power; and with power comes the seductive temptation to better one's condition at the expense of others.

Many asked me whether I had felt threatened or in danger at any point along the way. Other than a few close encounters with some fascinating wildlife, I could not think of a single moment when I felt uncomfortable or exposed.

Although I ended up travelling by myself, I never felt truly alone.

My vulnerability as a solo traveller seemed to serve as a bridge, making me more approachable. Whenever I would pull over, locals invariably came towards the bike and gave me a warm welcome with open smiles and easy laughter. In the end, it was not the wildlife or the unknown that posed a threat along this journey. It was the risk of closing myself off from the beauty and generosity of the human spirit that would have been the true danger.

I'm already putting together a rough plan for my next motorcycle journey. Perhaps riding across Asia on something a little less reliable. I will forever remember the lesson I learned along the way: the real adventures in life only begin when shit goes wrong.

***

On my final day in Cape Town, after a frantic morning of rushing around and tying up loose ends, I finally entered the airport and prepared to depart. As I approached the check-in desk, I nervously thumbed through the documents I needed to proceed, fully aware that my next few steps would bring closure to a profound chapter of my life. The staff member on duty inspected my passport, examining each of the many stamps I had acquired over the previous year and a half. A warm smile spread across her face.

"Mr. Anderson, you have been to so many countries. But how?" she asked.

"On a motorcycle," I responded. "All the way from Ireland."

"What an amazing adventure! How was it?"

"Wife-changing," I replied.

She chuckled at my comment before correcting me: "Surely you mean life-changing?"

"Yeah, that too."

She gave me a quizzical look as she handed back my documents. "Have a pleasant flight, Mr. Anderson."

***

Personal Diary Entry: 20th November 2014

Like the river, I was formed when the rains of my predecessors fell upon the distant mountains of my birth. I am not the first of my kind, nor will I be the last; I am simply the current embodiment of a timeless energy that has always been. From those lofty peaks, I began my journey, innocently seeking the path of least resistance, tumbling forwards, gaining momentum, my unique course influenced by the tributaries flowing into my being – family, friends, teachers, lovers.

As I matured, I grew, ever so slowly carving my own influence into the world around me. At times, I met with obstacles that threatened to impede my progress. I responded with stubborn turbulence until I eventually found a way through; and always, I was drawn relentlessly onwards by the gravity of time. I've experienced lakes of tranquillity where life seemed easy and peaceful, and

I've encountered tumultuous rapids of confusion and rage where the horizon ahead would disappear and the path forwards was not always clear. Amidst the uncertainty, I knew I would always prevail . . .

One day, I will reach my ocean and all that I am will dissipate into a boundless sea of possibility, eventually contributing to yet another cycle, another journey, another adventure.

Cork to Cape by the Numbers

549: days on the road, 18.04 months or 1.504 years.

62,840: kilometres ridden, or 39,047 miles.

3,000: approximate litres of fuel used, equivalent to 659 UK gallons or 792 US gallons.

1: number of crashes; this occurred in Nairobi, Kenya after swerving to avoid hitting the rear end of a matatu.

3: number of punctures; shockingly all from the roads in Kenya, two in the front, one in the rear.

9: number of tyres used along the way, 4 on the front and 5 on the rear.

40: number of countries visited.

7: number of ferries taken.

18: number of times I dropped my bike, as a result of deep sand, slick mud, or fatigue.

6: number of times I got pulled over by the police.

40–50: number of times I didn't pull over when flagged down by the police.

1: number of speeding tickets received.

33: number of books read along the way.*

*Reading is my refuge. The practice of transporting oneself to fantastical worlds, travelling through space and time, meeting the heroes and villains of history because of a specific arrangement of ink marks on a page will never cease to delight me. Throughout this journey, I always travelled with two books – one that I was currently reading and another that I had finished but was willing to trade for fresh material. Meeting new people always held a magical opportunity for experiencing an unread book. I had little choice in what was available and I discovered many hidden gems but also a few disappointments. I read the *Game of Thrones* saga in completely the wrong order. I found stories relating to the regions I was travelling through which added a depth and detail I otherwise might have missed. I dreamed of writing my own book, of how it might, one day, be stuffed into a stranger's backpack on its own journey.

About the Author

David "Irish" Anderson has been an outdoor educator and adventure guide for over thirty years. Originally from Ireland, he has lived and worked on every continent, including Antarctica, while pursuing his passion for exploration and discovery. He caught the motorcycle travel bug at a young age and has spent more nights sleeping under the stars than he can possibly remember. Irish currently lives in Queenstown, New Zealand, where he divides his time between guiding, writing, and planning more shenanigans.

Acknowledgements

This book is the culmination of an incredible journey, and it would not have been possible without the support and encouragement of many wonderful people.

First and foremost, I extend my deepest gratitude to my family whose unwavering love and support have been the foundation of my strength. To my parents, thank you for instilling in me the spirit of adventure and the courage to chase my dreams. To my siblings, Mel, Michael, Paul, Mary, Joan, and Gerrie, your constant encouragement and faith in me have been an invaluable guiding light. To my wonderful partner, Anouva, your love, patience, and understanding provided me with the sanctuary I needed to finally complete this book.

A special thanks to my editor, Elina Oliferovskiy, whose keen eye and attention to detail helped shape this book into what it is today. Thank you for believing in me, even when I didn't. You helped me find my voice as we brought this story to life. Your passion and dedication to your craft are deeply appreciated. I feel as though this book is as much yours as it is mine.

I am profoundly grateful to the countless individuals I met along the way. Your kindness, hospitality, and stories enriched my journey beyond measure. From the helpful stranger in the remote village to the fellow traveller who shared their wisdom, each of you played a special role in shaping this adventure: Sarah Laisney, Banu Kayali,

Kati Ox, Louis Tang, Katherine Tucker, Renske Hagg, Leyla Ahmet Meredith, Ruan Viljoen, Ingrid de Mattos, Sara El-Deeb, Tanja Lage, Naomi Viccars, Tomas Mossman, LivMa'ayan Knightly, Piers Pirow, Romain Mari, Lucia Giuliani, Alexander Honeybourne, Jane Bowen, Sina Klein, Judith Hannah, Axel Oz, Keri Redanz, Laura Sterner, Ed Peeters, Kihara Esther Wangui, Lovemore Kalinda, Ondine Bregeault, Shannon Murphy, Bushra Sultana, Kevin Convard, Silas Kisambira, Hamish Guthrie, Diana Harl, Ruaridh Stevenson, Amy Fraser, Fleur Williamson, Tozi Muringa, Hamish McMaster, Camilla Snugerrud, Maddie Leslie, Linnéa Maria Larson, Vicky Toria, Anselmo Blake, Mattia De Biasi, Flore Millet, Wubalem Tedla, and Joan MacLachlan.

To the motorcycle and overlander community, thank you for your camaraderie and your shared passion for the open road. Your tips, tricks, and tales will live with me for a lifetime: Damien Coke, Maya Mufti, Bishr Sakkal, Rob & Hanlie Reinecke, Ahmed Omran, Alpaslan Sramiş, Ross Clarke, Marith Spencer, Duesi Nator, Leonie Sinnige, Peter Scheltens, Brett Cooper, Francis M Muiruri, Alexander Conrad, Maurice Raleigh, Polly Marinova, Mouton van Zyl, Doug Wothke, Patrick Fuchs, Coleman Murphy, Kim Shaw, Gehad El Tehety, Nabil Gayar, Haytham El-Shayeb, Polytimi Boznou, Teresa Madariaga, Kfir Sharon, Ziggy Kiss, and Sam Watson.

To my friends and extended family back home and around the world, thank you for your unwavering support and for keeping me grounded. Your messages, calls, and letters were beacons of hope during the darkest times on the road: Gerard McElroy, Aiden McGrath, Harry Mac Intyre, Kyle Spinney, Danly Pessoa Calvo, Lynn Maclachlan, Brendan Anderson, Randi Kepecs, Carol Bradbury, Katie Anderson, Marly & Stevie Glasgow, Katie Allan, Niamh Duffy, Holly Santos Willmott, Pearse Anderson, Julie 'Juba' Christensen Pinto, Josh Anderson, Sean McElroy, Sandy Frizzell, Niall Duffy, Tony Barnett, Vanessa Anderson, Pete Meredith, Noa Bodner, Braden Tuohey, Randall Frizzell, Riaan Meintjes, Nick March, Shane Óg McNeill, Thomas & Kate Oxtoby, Sean Anderson, Sharron Grant, Chris Russell, Callum Foster, Kevin Anderson, Paul Shackleford, Claudia Senn, Kay & Dave Atherton,

Callum Foster, Terry Shearn, Molly Fisk, Colin Carpenter, Suzanne Wiley Pippen, Peter Artner, Michael Graney, and Mandela Leola van Eeden.

To Sarah, thank you for the moments we shared on this journey. Though our paths have diverged, the experiences and lessons we encountered are forever etched in my heart. Thank you for enabling me to understand the true meaning and power of forgiveness.

To the incredible team at Pure Ink Press – Leila Summers, Rebecca Bruckenstein, and my meticulous copyeditor, Lynn Thompson – thank you for transforming this book from a mere dream into a tangible reality. Your hard work and keen insights have brought this story to life in ways I could never have imagined.

Finally, to you, the reader: thank you for taking the time to immerse yourself in my story. Your willingness to share in my experience means the world to me. I hope this book resonates with you and inspires your own adventures.

With gratitude and hope,

David "Irish" Anderson

Glossary

Adhan is the Muslim call to prayer used to announce the time for one of the five daily prayers.

Arak is a distilled Levantine spirit of the anise family of drinks.

Auld sod is a reference to the old country, Ireland; my place of origin.

Benzine is a colloquialism for petrol, commonly used in some African countries.

C'est la vie is a French phrase that translates to "that's life" or "such is life."

Chadur is a full-body-length garment that covers the hair and body. It is worn in public by some Muslim women.

Dhow is a larger relative of the felucca, a sailboat primarily used for trading and longer voyages.

Doolally is a slang term for "losing one's mind" or "going crazy."

Expat is a term used to describe a person who lives outside their native country.

Felucca is a small boat propelled by lateen sails, oars or both, used on the Nile River.

Hijab is a head covering worn in public by some Muslim women.

Inshallah is an Arabic language expression meaning "if God wills" or "God willing" often used when discussing events in the future.

Lederhosen are leather breeches worn by men in Bavaria.

Mzungu is a Swahili word that literally means "wanderer" originally pertaining to the first Europeans to visit Africa. The term is now commonly used to refer to someone with white skin.

Pear-shaped is a colloquial phrase, chiefly British, meaning that something has gone wrong or awry.

Pension is a type of guest house, the term is typically used in Continental Europe.

Über is a German language word meaning "over" or "above."

Wadi Gnai is a small oasis about thirty minutes from Dahab, Egypt.

Bibliography

CHAPTER ONE

[1] Salles, Walter, dir. *The Motorcycle Diaries*; Buena Vista International (Latin America), Focus Features (United States), Pathé Distribution (United Kingdom and France), Constantin Film (Germany), 2004, Film

CHAPTER SEVEN

[2] The Walt Disney Company. "Conquer the Mountain – and a Mythic Monster." Accessed July 11, 2024. https://disneyland.disney.go.com/attractions/disneyland/matterhorn-bobsleds/.

[3] Zermatt Tourism. "Matterhorn. Come for the View. Stay for the Magic." Accessed July 11, 2024. https://www.zermatt.ch/en/matterhorn.

CHAPTER EIGHT

[4] Gibraltar Travel. "World War II Tunnels." Accessed August 19, 2024. https://gibraltar.com/en/travel/see-and-do/history-and-heritage/world-war-2-tunnels.php.

CHAPTER NINE

[5] Soligo, Marta. "Urban Social Life on Hold: Italian Communities and COVID-19." American Sociological Association Cultural

Section. August 14, 2020. https://asaculturesection.org/2020/08/14/
urban-social-life-on-hold-italian-communities-and-covid-19/.

CHAPTER TEN

6 *Encyclopaedia Britannica Online.* s.v. "San Marino."
Accessed August 19, 2024. https://www.britannica.com/place/
San-Marino-republic-Europe.

CHAPTER TWELVE

7 Al-Bayati, Sundus. "A City That Doesn't Forget: Sarajevo Thirty
Years after the War," *SAH Blog*, Society of Architectural Historians,
July 8, 2022. https://www.sah.org/publications/sah-blog/blog-detail/
sah-blog/2022/07/08/a-city-that-doesn-t-forget-sarajevo-thirty-years-
after-the-war.

8 *Encyclopaedia Britannica Online.* s.v. "Durmitor." Accessed July 11,
2024. https://www.britannica.com/place/Durmitor.

CHAPTER THIRTEEN

9 *Encyclopaedia Britannica Online.* History – North Macedonia.
"The Ottoman Empire," by Lorging Danforth. Accessed August 19,
2024. https://www.britannica.com/place/North-Macedonia/
The-Ottoman-Empire.

10 *Wikipedia.* s.v. "Millennium Cross."Last modified July 26, 2024.
https://en.wikipedia.org/wiki/Millennium_Cross.

CHAPTER FOURTEEN

11 Statista. "Population of Istanbul in Turkey from 2007 to
2023." Accessed August 19, 2024. https://www.statista.com/
statistics/899051/turkey-population-of-istanbul/.

12 "Constantinople." History.com. December 6, 2017. https://www.
history.com/topics/middle-east/constantinople#istanbul.

CHAPTER FIFTEEN

13 *Encyclopaedia Britannica Online.* s.v. "Parthenon." Accessed
August 19, 2024, https://www.britannica.com/topic/Parthenon.

CHAPTER NINETEEN

[14] "Monastery." Friends of Mount Sinai Monastery. Accessed July 11, 2024. https://www.mountsinaimonastery.org/monastery#overview.

CHAPTER TWENTY-TWO

[15] Bousquet, Delphine. "Understanding the Role of a Fixer." International Journalists' Network (2024). https://ijnet.org/en/story/understanding-role-fixer.

[16] Konstantinovsky, Michelle. "Why Is the Tropic of Cancer Important?" *HowStuffWorks* (InfoSpace Holdings LLC), November 30, 2023. https://science.howstuffworks.com/environmental/earth/geophysics/tropic-of-cancer.htm.

[17] *Encyclopaedia Britannica Online.* Companion – Science & Tech. "Why Does the Tropic of Cancer's Location on Earth Move Over Time?" by John P. Rafferty. Accessed July 11, 2024. https://www.britannica.com/story/why-does-the-tropic-of-cancers-location-on-earth-move-over-time.

[18] United Nations Children's Fund (UNICEF). Female Genital Mutilation in Sudan. Sudan: UNICEF. Accessed August 19, 2024. https://www.unicef.org/sudan/media/9386/file/FGM%20Factsheet-FINAL.pdf.

CHAPTER TWENTY-THREE

[19] Fleming, Lucy. "Ethiopia: The country where a year lasts 13 months." BBC News (2021). https://www.bbc.com/news/world-africa-57443424.

[20] Johnson, Erskine. "In Hollywood." *Dunkirk Evening Observer,* October 20, 1949. NewspaperArchive.

CHAPTER TWENTY-SEVEN

[21] "Believed to be the source of Ebola, this Kenya cave could be ground zero for the next pandemic." *Deccan Herald,* April 23, 2024. https://www.deccanherald.com/world/

believed-to-be-the-source-of-ebola-this-kenya-cave-could-be-ground-zero-for-the-next-pandemic-2990212.

22 Kron, Josh. "A Source of Thrills." *New York Times,* September 13, 2012. https://www.nytimes.com/2012/09/14/sports/on-the-nile-in-uganda-the-source-of-kayakers-biggest-thrills.html.

CHAPTER TWENTY-EIGHT
23 *Encyclopaedia Britannica Online.* s.v. "Idi Amin." Accessed July 11, 2024. https://www.britannica.com/biography/Idi-Amin.

CHAPTER TWENTY-NINE
24 Coetzee, Hendri. *Living the Best Day Ever.* Massachusetts: Digital on Demand, 2013.

25 Tsavo Trust. "Is there a lot of elephant poaching in Tanzania?" Accessed August 19, 2024. https://tsavotrust.org/elephant-conservation-in-east-africa-has-tanzania-got-a-handle-on-elephant-poaching/.

26 Okari, Dennis. "Mpeketoni Attacks: Four Possibilities." *BBC News,* June 17, 2014. https://www.bbc.com/news/world-africa-27890084.

CHAPTER THIRTY-ONE
27 McGregor, Ewan and Charley Boorman, dir. *The Long Way Down.* Aired October 28 - December 2, 2007, on BBC Two.

CHAPTER THIRTY-FIVE
28 *Encyclopaedia Britannica Online.* "Big Hole," by John P. Rafferty. Accessed July 11, 2024. https://www.britannica.com/topic/Big-Hole.

29 McKechnie, W.F. "Diamond Exploration and Mining in Southern Africa: Some Thoughts on Past, Current, and Possible Future Trends." *The Southern African Institute of Mining and Metallurgy* 119, no. 2 (February 2019): 124. https://www.saimm.co.za/FullJournal/SAIMM-201902-Feb.pdf.

30 Smith, David. "South Africa's Jacob Zuma Promises to Take HIV Test in AIDS Policy Reversal." *The Guardian.* December 1, 2009. https://www.theguardian.com/world/2009/dec/01/jacob-zuma-hiv-test.

31 McGreal, Chris. "South Africa Ends Long Denial Over AIDS Crisis." *The Guardian,* November 30, 2006. https://www.theguardian.com/world/2006/nov/30/southafrica.aids.

32 Welsh, Frank. *South Africa: A Narrative History.* First Edition. New York: Kodansha America Inc., 1998.

33 Sabino, Fernando Tavares. *O Tabuleiro de Damas (The Checkerboard).* Brazil: Record, 1988; Quote Investigator®. "Quote Origin: Everything Will Be OK in the End. If It's Not OK It's Not the End." Accessed July 11, 2024. https://quoteinvestigator.com/2023/01/01/everything-ok/.